Overstep Revelationism Contrivance in Ether Synaptemics

Victor

Overstep Revelationism Contrivance in Ether Synaptemics

First Edition September 2025

Written by Victor

Contents

Chapter 1

CHAPTER 1: INTRODUCTION

1.1. Background

Modern enterprises have started moving their services to the cloud computing to reduce their operational costs. The cloud services like Amazon web services, Azure, Google cloud are some of the services that provide complete solutions for hardware or software provisioning using the administrative interface provided by the Cloud Service Provider (CSP). However, once an organization moves its data to the cloud, it loses the control on its data with respect to location [1]. This gives rise to the issue of regulations and compliance. For example, US federal data cannot be stored on the servers which are placed outside US. This becomes a limitation of the cloud [2].

Larger organizations which are spread across the globe are moving to the cloud at a slower pace due to the compliance and regulatory issues that may arise due to the different compliance rules in different countries. Certain industries like financial, federal etc. are still hesitant to move their critical data to the cloud due to compliance as well as security uncertainty. NIST's definition of cloud computing highlights the features of the cloud but does not mention the security features that should be adopted by the cloud [3]. The absence of security framework for the cloud gives rise to the vulnerabilities which can be exploited by attackers to steal the critical data from the cloud [4].

Cloud Security Alliance (CSA) which focusses on the cloud security has come up with some best practices for the cloud environment. At the same time CSA analyses the types of threats that are faced by the cloud users. CSA has listed top 12 threats to the cloud computing which also include the threats like DOS and cloud service abuse from malicious inside users [5]. To detect such threats, we need to deploy the prevention or at least a detection technique to safeguard the data and to gain the confidence of existing and potential customers. Firewall is one such technique that allows to block or allow the network packets to enter the cloud based on the rules configured on firewall. The main drawback of the firewall is that it cannot detect zero-day attacks as well as the insider attacks [6]. A more advanced device i.e Network Intrusion Detection system (NIDS) is used to detect the malicious packets entering the cloud. Such a device can detect the intrusions and raise an alarm if any intrusion packet is detected on the network. It has an advantage of detecting zero-day attacks also (If it is anomaly based).

This work was focused on anomaly-based NIDS and addressed the key issue of accuracy and false alarms. The aim of this work was also to improve the performance of NIDS by enabling it to make the decisions faster by inspecting fewer features of a network packet. The proposed NIDS can be useful in the cloud environment where a large volume of packet needs to be processed by NIDS.

1.2. Cloud Computing

With the rapid development of storage and processing technologies as well as increase in the popularity of internet, computing resources have become more affordable and easily accessible. This has given birth to a new computing model known as cloud computing as shown in figure 1.1. In this model the resources like processing power, storage, memory etc., are treated as utilities and are leased by the users as per their need.

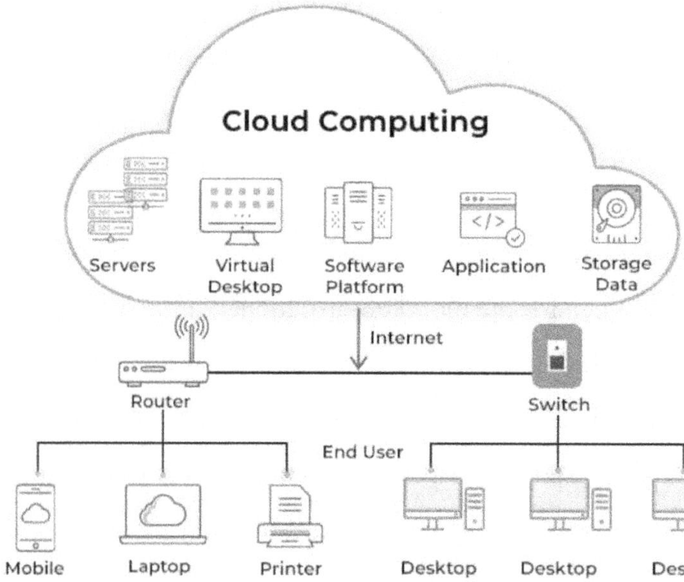

Fig.1.1. Cloud Computing Model

The resources are released when it is no longer needed [7]. Figure 1.1 shows the cloud computing model with various resource types and the heterogenous clients.

NIST defines the cloud as "*Cloud computing is a model for enabling convenient, on-demand network access to a shared pool of configurable computing resources (e.g., networks, servers, storage, applications, and services) that can be rapidly provisioned and released with minimal management effort or service provider interaction*" [3].

Cloud computing is not a new technology, but it is the new operating model that is comprised of many existing technologies to run business in a different way. Virtualization can be considered as one of the major bricks of the cloud computing [8]. Some of the features that make cloud computing an attractive choice for organizations is shown in the figure 1.2. Two important enabling technologies that make the cloud computing a success among the organizations are virtualization and multi-tenancy [9]. Virtualization is a concept that is used to hide the physical details of the hardware resource like disk, CPU etc. and provide virtualized resources to the applications. For cloud computing this helps to optimize the cost by creating multiple Virtual Machines (VMs) from a single physical machine called as a host [10]. Mult-tenancy feature allows the CSP to host multiple tenants or organizations on a single physical resource using virtualization.

1.2.1. Types of the Cloud

Depending upon the deployment models, clouds can be categorized as public cloud, private cloud, and hybrid cloud.

a) Public Cloud

Public cloud makes the resources available to the common user or an organization over the public network [11]. However, in this case the user cannot manage the cloud infrastructure but can utilize the limited services that are provided by CSP. In most of the cases public clouds do not charge or may charge based on the usage of services. However, public clouds provide a limited set of services to the end users. This type of the cloud may be owned and managed by an enterprise, government, or any academic institute. Examples of such cloud services include google drive, Dropbox etc. One of the shortcomings of the public clouds is that they lack the fine-grained control over data or security. Hence security is a major concern in public clouds [12].

b) Private Cloud

Private cloud is dedicated to a single enterprise or any department of an organization. It provides the service to a single organization or a department. The services can include application, processing power, memory, security etc. Private cloud is managed by the organization itself or may be outsourced. It may exist in the enterprise premises or out of it. Private cloud provides the control over the organization data as well as highest level of security and performance [13]. However, setting up the private cloud increase the infrastructure cost of an organization. It also needs expertise to set up this model.

c) Hybrid Cloud

Hybrid cloud is a combination of public and private clouds. It combine the benefits of both the types of clouds. In this type of cloud, a part of the infrastructure is placed on the public cloud and another set of infrastructure is maintained by the organization himself. It is mostly used when there is some critical data that needs a tighter control as well as security. Such type of data is placed in the private cloud and rest of the data can be moved to the public cloud. However, designing such cloud is a challenge as it requires the identification of the cloud components that needs to be split between private and public clouds. Table 1.1 gives the comparison of the three types of clouds.

Table 1.1. Comparison of Cloud types on various Parameters

Parameter	Public Cloud	Private Cloud	Hybrid Cloud
Cost	Lowest	Highest	Medium
Security	Lowest	Highest	Medium
Effort	Lowest	Medium	Highest
Location	CSP	Enterprise	CSP and Enterprise
Control	Least	Most	Medium

1.2.2. Features of Cloud Computing

Cloud computing has been emerging as a popular information technology platform for many organizations. Such Organizations have migrated their services like email, database, virtual machines to the cloud. This allows organizations to save infrastructure cost as well as the operational cost. Some of the features of cloud computing that benefit an organization are shown in figure 1.2

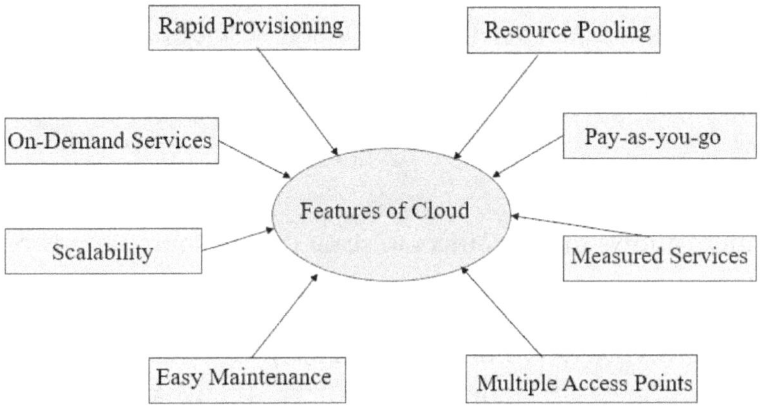

Fig.1.2. Features of Cloud Computing

a) On-Demand Services

One of the most interesting and useful features of cloud computing is that consumer organizations can provision the resources themselves as and when needed. CSP provides an interface to the consumer organizations to provision or deprovision the resources like VMs, processing power, storage, firewall, load balancer etc. The provisioning of these resources can be automated by using the templates that are provided by CSP. This results in the efficient utilization of computing resources.

a) Resource Pooling

Major reason for the lower cost of the cloud services is the fact that CSP uses the pooled enterprise resources to share among multiple customers. VMs is the building block of resource pooling [14]. Using this technology multiple VMs can be provisioned on a single physical host. These VMs can then be assigned to different customers using multi-tenant concept. However, having multiple VMs on a single host also poses some security challenges. In case a physical host is compromised, all the virtual machines which are managed by it may also get compromised. Also, any attack at the hardware level can bring down all the virtual machines, thus impacting multiple customers at the same time. Thus, resource pooling also results in some security vulnerabilities.

b) Multiple end Points

Cloud services can be accessed using heterogenous clients which include traditional desktops, mobile phones, iPads etc. This makes accessing the cloud services much easier and location independent. The cloud services are accessed through the public network, thus making the resource access most cost effective and economically viable.

c) Elasticity

This feature allows the customers to scale their resources up vertically or horizontally. The resources can also be released when no longer needed. For example, a customer can provision extra storage to the existing VM when needed and deprovision it when no longer needed. Customer will be charged only for the duration he used the extra storage. The billing stops as soon as he de-provisions the extra storage. This feature is useful when customer requirements change over the time. This helps to save the cost [15].

d) Metered Charges for Resource Utilization

Metered charges, also called the pay-as-you-go allows the customers to pay for only the services that they have used and for the duration that they have used these services. CSP has the capability to monitor the resource usage of the services by the customer. This also is an effective way to save the cost. For example, in Azure cloud one can provision the VMs, but he will be charged only when those VMs are up and running. Once the VMs are shut down, the billing stops. Thus, even though one provisions the resource, he is billed only when resource is used. This is an advantage for end users and customer organizations [16].

e) Rapid Provisioning

CSP provides easy-to-use administrative interface to the customers to provision and manage the resources. Through this interface it is easier and quicker to provision these resources. For example, Azure provides administrative console to manage VMs. The task of provisioning becomes more quicker using ready to use templates provided by CSP [17]. Such templates can be used to provision VMs with already defined configurations.

1.2.3. Architecture of Cloud Computing

Based on the service model, cloud can be divided into three layers i.e Infrastructure layer, Platform layer and Application layer as shown in figure 1.3. Each layer of the cloud provides all the services which are offered by the layer below it and some additional services to the customers [18]. The services that are provided through these layers are termed as Infrastructure as a Service (IaaS), Platform as a Service (PaaS) and Software as a Service (SaaS). As we move up the layers from IaaS to PaaS, the responsibility of CSP for managing the cloud infrastructure increases whereas for the customer it decreases. For example, the responsibility of CSP is least at the IaaS layer and most at the SaaS layers. Similarly, customer has the maximum responsibility at SaaS layer and least at IaaS layer. From the security perspective, CSP must take care of all the security measures at SaaS level.

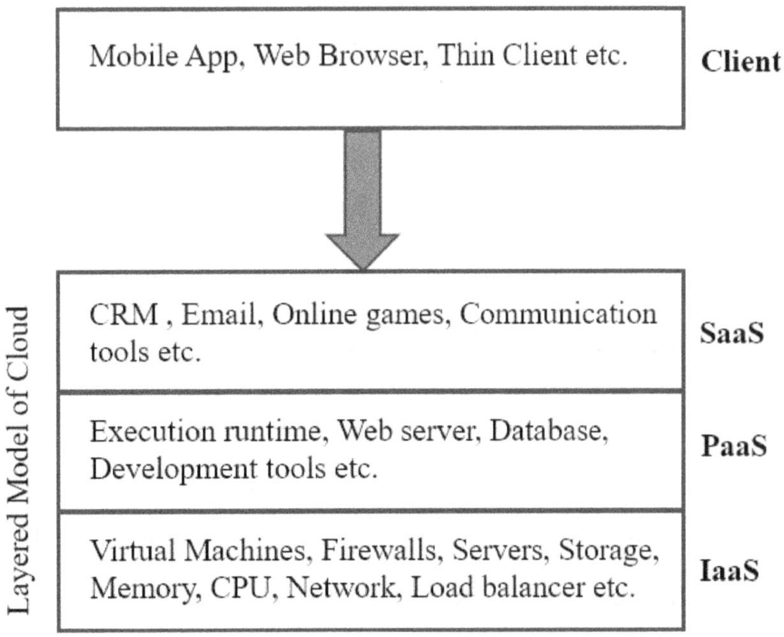

Fig.1.3. Service Models of Cloud Computing

a) Infrastructure as a Service

This layer provides the virtualized resources like CPU, storage etc. that are accessible through internet. The main technology used at this layer is virtualization. A physical resource is partitioned into multiple virtual resources which can then be provisioned on-demand by the customers. Usually customers have full control over the resources at this layer [18]. Microsoft Azure, Amazon C2 are some of the widely used IaaS models.

b) Platform as a Service

PaaS exists on the top of IaaS layer. It provides all the services which are provided by IaaS as well as the application framework to deploy and run the applications [19]. This layer helps the developers and DevOps teams to deploy their applications on the cloud without being concerned about the underlying hardware or application framework. In this model CSP is responsible for managing the IaaS layer. Google AppEngine, Amazon Simple DB are some of the popular PaaS [20].

c) Software as a Service

SaaS layer sits on the top of PaaS and provides all the services of PaaS along with applications which can be accessed through the internet. The applications at this layer are configured for high availability and can be accessed through internet. The underlying services at IaaS and PaaS are taken care by the CSP whereas customer has the control of SaaS. The applications are usually provided on pay-as-you-go or subscription basis [21].

1.2.4. Security Threats in Cloud Computing

It may sound that when an organization moves its services to the cloud, it is assured of the security of its data. However, cloud brings some risks with itself. First and foremost, problem faced by customers is the lack of control on their data [6]. Customers are unaware of the location of the server on which their data is hosted. Data by CSP is maintained in data centers which are geographically distributed. Sometimes multiple copies of the customer data are maintained at multiple locations for high availability [22]. In such cases, the traditional security systems like firewalls, host-based intrusion detection systems do not provide adequate security to the customer data. Cyber- attacks can occur at any of the three layers of the cloud and can impact VMs, hypervisor, Applications or Network. Hypervisor is a tool that runs on the host and is used to manage multiple VMs [23]. Figure 1.4 shows the broad categorization of attacks and table 1.2 lists the specific attacks under these categories.

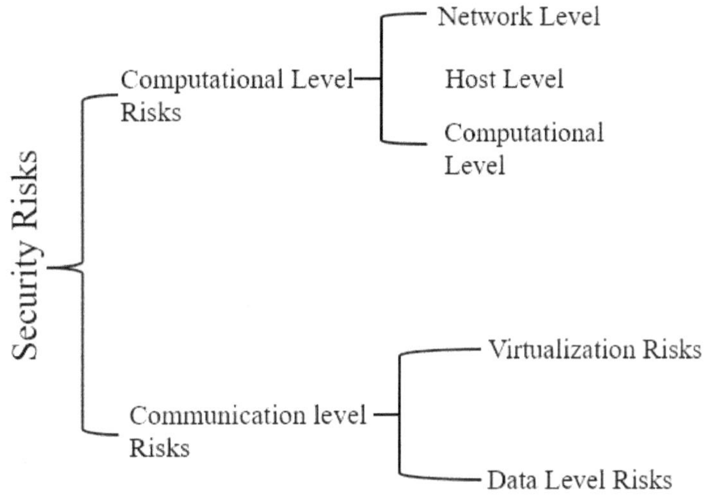

Fig.1.4. Categories of Cyber Attacks

Table 1.2. Possible Cyber-attacks on Cloud Computing

Category of Attack	Description
Network Attack	Sniffing, BGP Attacks, DNS Attacks, Reused IP Addressing etc. DOS, DDOS, Cookie Poisoning, Dictionary Attack, SQL Injection, Trojan Horse, etc.
Aplication Level Attacks	Password cracking, Profiling, DOS, Unauthorized access, Virus, Worms etc.
Host Level Attacks	VM Cloning, VM Migration, VM escape, VM Sprawl, VM hopping, VM poaching.
Virtualization Hypervisor Level Attacks	VM-to-VM attack, issues with VM introspection, Hyperjacking ,
Data level challenges	Data Segregation, Data Isolation, Data lineage, Data leakage , CIA of data.

1.3. Need for Intrusion Detection System in Cloud Computing

Due to the large surface area of the cloud and multi-tenancy nature, any attack on the cloud can cause damage at a large scale. The most critical attack vector of cloud computing is resource sharing which means that multiple VMs hosting the data of multiple organization are managed by a single physical host. Thus, they share the same underlying physical hardware.

Cloud computing has become an attractive choice for many organizations due to its lower cost as well as high availability. The features like metered charges, elasticity, easy provisioning, geographical replication and Content Delivery Network (CDN) allows an organization to reduce the downtime as well as cost and increase the performance [24]. Metered charges allows an organization to pay for only the service based on their usage, Elasticity makes it possible to scale up or scale down the resources as per their need. The cloud interface allows an organization to provision the resources without caring about the underlying hardware. For example, a cloud would provide an interface to configure a routing table without going into the details of the underlying router configuration. Data replication across the geographical locations provides the high availability. CDN allows the data to be brought closer to the user, thus increasing the performance. Users can access the contents from their local CDN server more easily.

The openness and the amount of the critical data that is stored in the cloud makes it a target for the intruders. The dispersed nature of the cloud and the large number of the endpoints make it vulnerable to the cyber-attacks. Due to the large surface area of the cloud and multi-tenancy nature, any attack on the cloud can cause damage at a large scale. For example, an attack like side channel on the hypervisor can compromise multiple VMs which are controlled by it. Thus, we need to have a tool that can monitor the network traffic of the cloud and raise an alarm whenever a suspicious packer is detected [25]. This will help to detect and prevent the attacks in a timely manner.

The term "Intrusion" refers to unauthorized access to the data or network for malicious purposes. Since the cloud is a multi-tenant entity, the attack surface area becomes larger as compared to a single organization. Thus, the impact of an attack is manifold. Any such malicious activity on the cloud network needs to be detected immediately to protect the user data. This helps to ensure Confidentiality, Integrity, and Availability (CIA) of the resources.

An Intrusion Detection System (IDS) is a combination of hardware and software or a hosted service (in case of cloud) that monitors and analyzes the activity on the host or network and raises an alarm if any malicious activity on the network or on the host is detected. Machine Learning Techniques (MLTs) like Support Vector Machines (SVM), Artificial Neural Networks (ANN), Decision Trees (DT), Naïve Baye Algorithms (NBA) etc. are preferred choice for researchers to develop and evaluate IDS. To improve the performance of IDS, some researchers have proposed Deep learning techniques like multi-layer perceptron, Recursive Neural Network , Convolutional Neural Network, Auto-encoders [11],[12] etc. Autoencoders, which are unsupervised neural networks have been more specifically used in feature representation in combination with some classifiers [28]. Ensemble machine learning techniques have also been used to increase the performance of IDS [29]. Ensemble technique combines multiple machine learning algorithms to create a model that performs better than any of these individual models.

Volume of incoming data in the cloud is huge and is high dimensional in nature. In such cases, machine learning is one of the most popular techniques to build an IDS model that can predict with fewer features . ANNs, SVMs, NBA etc., have been successfully used by some researchers to detect the anomalies under such circumstances [30]. However, one of the drawbacks of IDS is that they tend to raise false alarms which needs human intervention. Hence accuracy is one of the main concerns for an IDS in the cloud. Second, the cloud is characterized by high volume of incoming data which is high dimensional in nature. Thus, IDS must process all the features of a packet to find the malicious packet, which is time consuming. This again raises an important question of how we can make an IDS to predict using fewer features, yet with higher accuracy. This research question has not been fully addressed yet. Research has been primarily done to increase the accuracy of IDS, but less effort has been made to optimize both accuracy as well as number of features at the same time.

IDS can be classified based on the data that they analyze to make the decision or based on the detection method that they use to make the decisions.

1.3.1. Components of an Intrusion Detection System

At the architectural level an IDS has four main components. i.e, Data collector, Data pre-processor, Detection module and Alert module as shown in figure 1.5 [31].

a) Data collector

Data collector module collects the data from the network or the host which it needs to monitor. It is a software agent that is installed on each host or network to be monitored.

b) Data Preprocessor

Data preprocessor may be a separate module or part of the data collector module. This module performs the preprocessing tasks like cleanup, normalization etc.

c) Detection Module

This module correlates the data and uses the detection algorithm to decide whether the intrusion has happened or not. This module may use misuse or anomaly detection algorithm for intrusion detection.

d) Alert Module

It works based on the decision taken by the detection module. Alert module raises an alarm if intrusion is detected by the detection module, otherwise no action is taken by the alert module.

In some cases, some of these components may reside on different hosts. Such IDS are called as distributed IDS [32]. Communication between modules is one of the biggest challenges in such IDS.

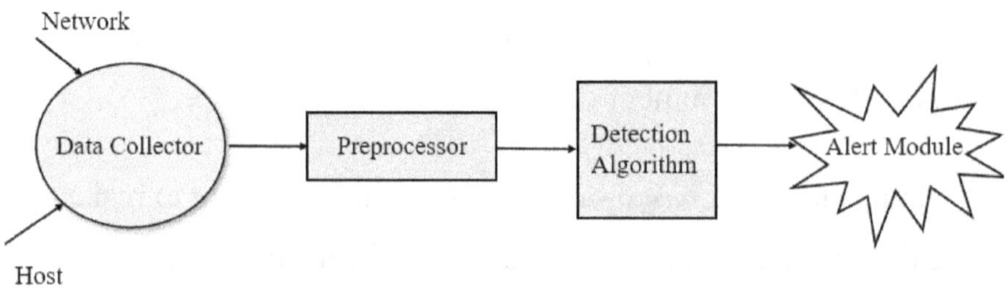

Fig. 1.5. Components of Intrusion Detection System

1.3.2. Classification of Intrusion Detection System Based on Data Source

Based on the source of data that IDS inspects to detect intrusions, IDS can be classified as Host Based IDS (HIDS), Network Based IDS (NIDS) and Hypervisor Based IDS (HyIDS). Each of these IDS uses different data source to detect intrusions.

a) Host Based Intrusion Detection System

HIDS inspects the data that originates from the physical resource or audit logs like server logs, router logs, database logs, operating system logs, firewall logs etc. These logs are analyzed to detect any anomaly which is then treated as an attack. This method can also be used to detect insider attacks that do not involve network traffic like unauthorized deletion or modification of a file [33]. This type of IDS needs to be installed on every host that needs to be monitored. However, with the increased number of hosts on a network like cloud, this type of IDS is not economically and technically feasible. Cloud infrastructure being geographically dispersed, it will need a more complex management process to manage the HIDS infrastructure. Fig.1.6. shows three hosts with HIDS installed on them.

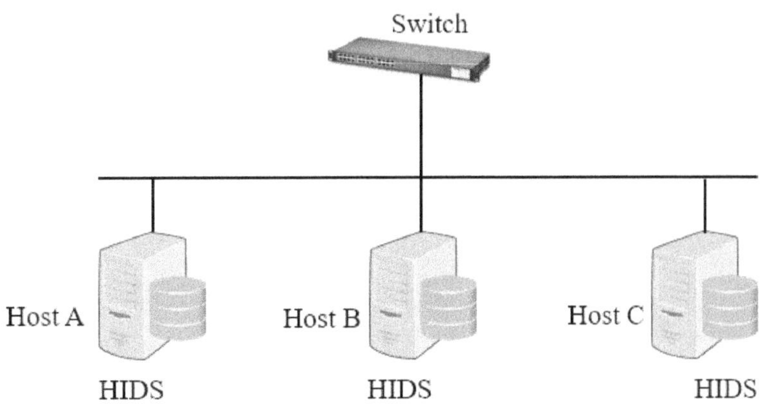

Fig.1.6. Host Based Network Intrusion Detection System

b) Network Based Intrusion Detection System

NIDS monitors and analyzes the network traffic captured from the network and detects any malicious packet in it. NIDS can analyze the overall traffic pattern or inspect at the protocol level using deep packet inspection to detect the abnormality in the network packet [34]. NIDS raises an alarm if any possible intrusion is detected. NIDS can be placed at the perimeter to detect external attacks and at the internal network segments to detect the internal attacks. However, in case of cloud computing NIDS should be able to analyze the incoming network traffic at a high speed without dropping any packets to prevent the intrusion. Snort is one of the popular examples of NIDS.

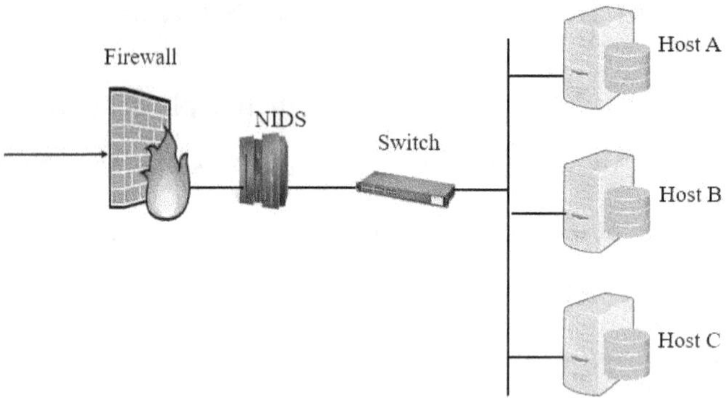

Fig.1.7. Network Based Intrusion Detection System

Figure 1.7 shows a NIDS placed behind a firewall to inspect incoming network packets before they enter local area network.

c) Hypervisor Based Intrusion Detection System

The key technology to enable the virtualization is hypervisor. It is a software that allows the creation and management of hardware resources among multiple VMs [35]. An example of a hypervisor with four VMs is shown in figure 1.8.

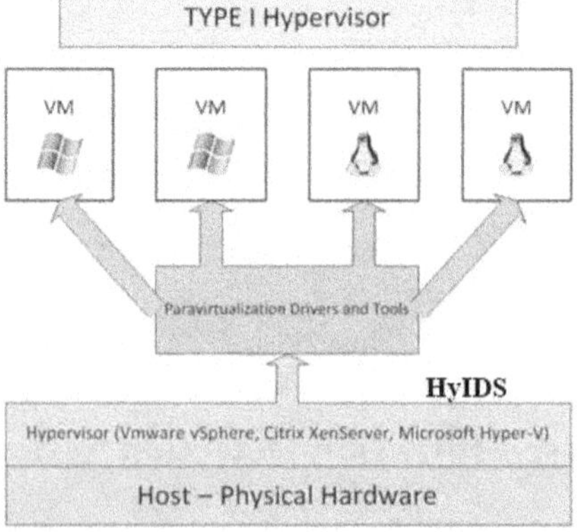

Fig.1.8. Hypervisor Based Intrusion Detection System

The use of hypervisor and VM technologies are also a security threat, as these technologies are vulnerable to VM level attacks. Vulnerabilities in VM infrastructure can be exploited by an attacker to launch attacks like DOS and DDOS. To prevent such attacks HyIDS are used. HyIDS monitors the traffic between guest and VMs and between the VMs as well [18],[36], thus detecting any anomaly in the network packets. HyIDS must be installed on each Hypervisor (Host machine) that needs to be monitored.

1.3.3. Classification of Network Intrusion Detection System Based on Detection Method

All types of IDS i.e. HIDS, NIDS and HyIDS use any of the two detection methods i.e Signature-Based method (*misuse method*) or Anomaly-Based method to detect the anomalies in the data or network traffic which are considered as possible attacks. Accordingly, the names are given to IDS based on their type and detection method that they use.

Detection methods refer to the algorithm that an IDS uses to detect the attacks. The efficiency of an IDS depends upon the type of detection algorithm that is used and the location where IDS is installed. For example, an IDS placed at the perimeter of the network won't be able to detect internal attacks. An example of such an IDS is a firewall placed at the perimeter of network.

Based on detection methodology, NIDS can be broadly classified as Signature based NIDS, Knowledge based NIDS and Anomaly based NIDS. NIDS that uses anomaly detection method is termed as Anomaly Based Network Intrusion Detection System (ANIDS) and the NIDS that uses misuse detection method is termed as Signature Based Network Intrusion Detection System (SNIDS). On the other hand, NIDS that uses expert knowledge is termed as Knowledge Based Network Intrusion Detection System (KNIDS). However, each of these types of NIDS uses the network traffic as a source to detect the intrusions.

In our work we have limited our scope to anomaly-based network intrusion detection system using supervised learning algorithms.

a) Signature Based Network Intrusion Detection System

A signature is a pattern or characteristic of a packet that classifies it as an attack. This pattern is used as the footprint of the attacker and helps to detect the intrusions. SNIDS are based on the principle of pattern matching. They use "*if-then*" rules for detection and can detect the known attacks only [33]. SNIDS requires a database of previously known attack signatures to match against. The pattern matching algorithm compares the incoming packets against the signatures in the database and raises an alarm if any match is found . Antivirus systems like McAfee is a typical example of SNIDS. Algorithm1.1 shows the typical misuse detection algorithm of SNIDS.

Algorithm 1.1. SNIDS Detection Algorithm

```
Input: signature database, incoming network packets
Output: malicious packet, alarm
m.db <- sdb                      //load SIDS database in memory
for  i = 1 to eo(m.db)             //Compare signature at ith
                                       //  record in database with
                                       // incoming  packet
if p[i] = m.db[i]
    mark the packet at i as anomaly.
  i <- i+1
end for
```

b) Knowledge Based Network Intrusion Detection System

KNIDS can also be called as the Expert Knowledge-Based NIDS. Firewall is an example of KNIDS. For these detection systems, an administrator needs to have the knowledge of attacks or the source of attacks that he wants to prevent. Administrators generally configure KNIDS to prevent such attacks. For example, administrators normally blocks ICMP ports to prevent DOS and DDOS attacks [37]. KNIDS generally allow or block certain type of network traffic based on rules. The traffic can be blocked based on source IP addresses, source port, domain name etc. For example, we can allow all the traffic to enter the local area network accept the one that originates from a certain IP address which we consider as a suspicious one. In more strict case, we block all the incoming traffic but allow only certain packets from selected hosts to enter the network.

c) Anomaly Based Network Intrusion Detection System

ANIDS are based on the principle that the normal behavior of the system can be differentiated from the abnormal behavior introduced by malicious activity. In terms of network traffic, ANIDS needs to be trained on the labelled data to create a predictive model. The labels can be "normal" for the normal traffic and "anomaly" for the suspicious packet. The trained model can then label the future data points in terms of "normal" or "anomaly". Such a model is created using statistical tools or MLTs. Some of the challenges that are faced by each of these models is shown in the table 1.3.

Table 1.3. Challenges faced by each type of Intrusion Detection System

IDS type	Challenges
SNIDS	✓ SNIDS database needs to be updated with newly detected attack signatures. ✓ SNIDS database grows with time as new attack signatures are detected. ✓ With increased number of records in SNIDS database, memory required load the database also increases. ✓ Major drawback of SNIDS is that it cannot detect the zero-day attacks.
ANIDS	✓ ANIDS have the tendency to raise false alarms which may need human intervention. ✓ Getting the sample Intrusion dataset to train the ANIDS model is a challenge. ✓ The word "normal behavior" can mean different for differenr environments.
KNIDS	✓ KNIDS requires administrators to have knowledge about source of attacks. ✓ Efficiency of KNIDS depends upon the knowledge of an administrator. ✓ KNIDS cannot detect zero-day attacks.

1.4. Importance of Machine Learning Techniques in Network Intrusion Detection Systems

Increase in the number of endpoints as well as the type of attacks and the volume of incoming traffic justifies the use of ANIDS in cloud computing [38]. MLTs are used in mining the network traffic to find a meaningful pattern in it. Any deviation from this pattern is flagged as an Intrusion. MLTs build an abstract model which defines the relationship between the network packet features and the class variable. This is done by training the specific machine learning algorithm on the network traffic and learn the pattern in the data. Since machine learning algorithms can handle large volume of data, they are suitable for developing ANIDS for cloud environments [39]. Machine learning techniques broadly fall under two categories viz. supervised and unsupervised techniques. This classification is based on the type of the learning used for training the models.

Supervised methods like SVM, ANN, NBA, DTs etc. use the classification method to assign the known labels to the instances. This is done by determining the relationship between each independent feature of the network traffic and the class variable [40]. The model is trained on the network traffic that contains "normal" as well as "anomaly" labels. These ANIDS models can monitor the network packets and detect similar attacks in the future. The drawback of the supervised model is that they require the dataset on which they can be trained. However, they give more accurate results than unsupervised methods. The accuracy of this type of ANIDS can be increased by training the model on heterogenous data [41]. A typical supervised method for classification is shown in the figure 1.9.

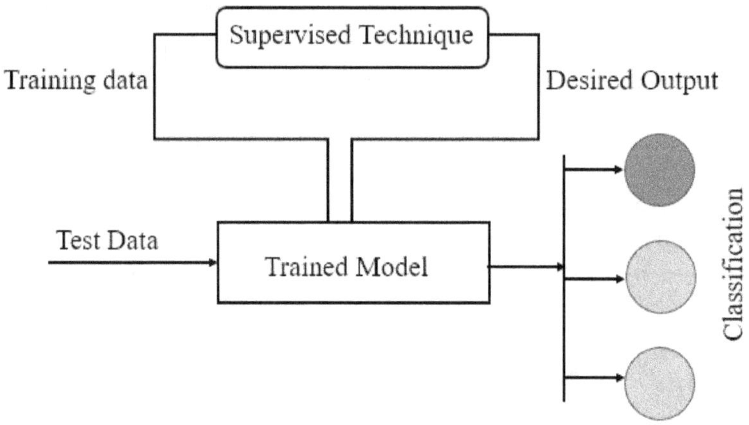

Fig. 1.9. Supervised Machine Learning Model

Unsupervised method divides the data into multiple disjoint groups (clusters) such that the union of these groups gives the original dataset. The members of a cluster are as similar as possible and bear no or less similarities from the members of other clusters [42]. To determine the membership of an instance, membership function like Euclidean distance function is used. K-means clustering, Fuzzy C-means clustering are some of the clustering methods. The cluster with more instances is treated as normal and cluster with less instances is treated as anomaly because anomalies occur fewer times than the normal pattern. The problem in this case is to decide on the number of K, especially when detecting the polynomial type of attacks. Unsupervised models also take more time to converge, and accuracy is less as compared to supervised models. The main advantage of such method is that the ANIDS using unsupervised method does not need training data. Figure 1.10 shows the classification of MLTs, and table 1.4 gives some of the major differences between supervised and unsupervised learning.

Table.1.4. Difference between Supervised and Unsupervised Learning

Supervised Learning	Unsupervised Learning
• Training dataset is labelled.	• Dataset is not labelled.
• It uses training dataset to train the model.	• It does not need any training dataset.
• Test dataset is classified based on the knowledge gained through training dataset.	• It uses the properties of the data to classify the dataset.
• Number of classes are known in advance.	• Number of classes are unknown which is a challenge.

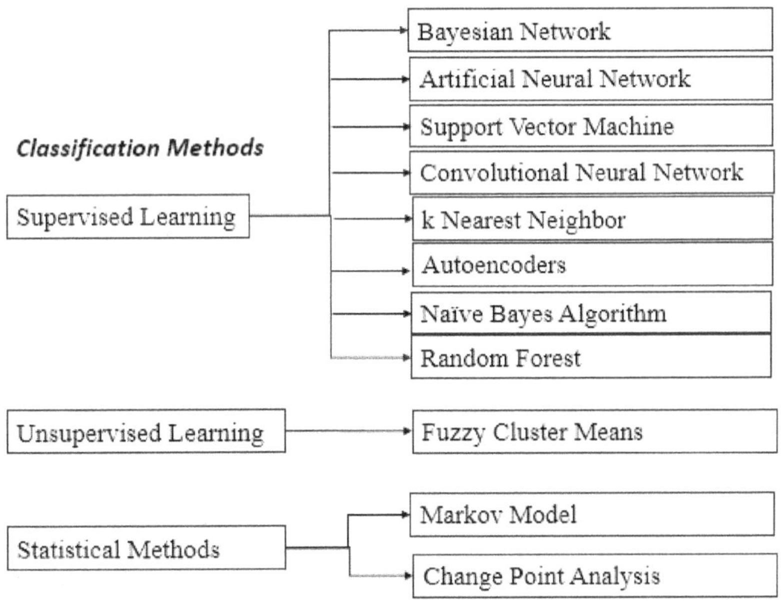

Fig.1.10. Classification of Machine Learning Techniques

There are other type of classifiers called One class classifier that have been successfully used in ANIDS. One class classifier are trained on the normal data only. Such models gain the capability of identifying the normal traffic and anything which is not considered as normal is flagged as an Intrusion [29]. However, One class classifiers have shown lesser accuracy as compared to supervised models [28].

1.4.1. Dimensionality Reduction using Machine Learning Techniques

Dimensionality reduction is a technique used to reduce the number of features in a dataset while retaining as much relevant information as possible. In other words, it is a process of transforming high-dimensional dataset into a lower-dimensional space that still preserves the essence of the original dataset [43].

In machine learning, high-dimensional data refers to data with many features or variables. The curse of dimensionality or overfitting is a common problem in machine learning, where the performance of the model deteriorates as the number of features increases [14]. This is because the complexity of the model increases with the number of features, and it becomes more difficult to find a good solution. Hence, high-dimensional data can also lead to overfitting, where the model fits the training data too closely and does not generalize well to new data [44]. Dimensionality reduction can help to mitigate these problems by reducing the complexity of the model and improving its generalization performance. There are two main approaches to dimensionality reduction i.e. feature selection techniques and feature extraction methods. Fig 1.11 shows the categorization of dimensionality reduction techniques. As shown in figure feature selection methods can further be classified as filter methods and wrapper methods.

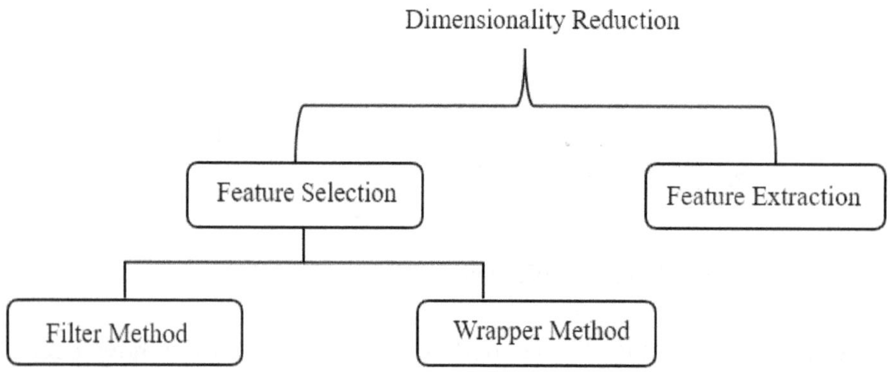

Fig.1.11. Classification of Dimensionality Reductin Techniques

a) Reducing the Noise in Data using Feature Selection Methods

With large number of features in network traffic, the performance an ANIDS can suffer. Some of the features in the network traffic may be irrelevant and contribute to the noise. Having large feature set can result in overfitting of the model and at same time increase the training as well as testing time [45]. This problem becomes more critical in case of cloud computing where incoming and outgoing traffic is huge. Thus, it is necessary to remove the redundant features from the sample dataset to optimize the model, save the computational cost as well as reduce the training and testing time. Eliminating the irrelevant or redundant features can also increase the accuracy of the model.

Feature selection is the process of evaluating a subset of features $\{x_1, x_2, x_3 \ldots\}$ from the feature space X_i and eliminating all those features that have no or have less predictive information for classification. Feature selection methods can broadly be classified as filter methods and wrapper methods.

Filter Method for Feature Selection

Filter methods use statistical techniques to calculate the correlation between each independent features and the class variable. Features are then ranked according to their correlation value. Higher correlation with the class variable means more predictive power. Thus, features with higher correlation value are selected for the classification model and those with no or low correlation with class variable are discarded [46]. A general correlation-based feature selection algorithm is shown in Algorithm 1.2. Determining the correction between each feature and the class variable is the basis for developing the abstract model for machine learning [47].

Algorithm 1.2. Filter Algorithm

```
Input : Dataset with N features, class feature y
Output:  Feature subset n , n<N
for i= 1 to N do
n[i] <- corr[Ni,y]            // calculate the correction
 i=i+1                        between feature Ni with class feature y
end for
Select top features from n[i]
```

Wrapper Method for Feature Selection

Wrapper method repeatedly evaluates a set of attributes on a classifier to assess their predictive performance. Wrapper method is more efficient as its takes into consideration the classifier hypothesis [46]. While filter method is independent of any classifier, wrapper method may give different results when different classifiers are used [33]. Wrapper method uses Greedy Search to search the feature space and select the features for the evaluation.

b) Importance of Feature Extraction in Anomaly Based Network Intrusion Detection Systems

One of the important characteristics of a good training dataset is that the independent features should have strong relationship with the class variable, while at the same time, the independent features themselves should not be collinear. To remove the collinearity between independent features, feature extraction technique is used. This effectively reduces the dimensions of data set.

Feature extraction transforms the data with high dimensions into a meaningful representation of the same data but with low dimensions by using Eigen values and Eigen vectors [46]. It combines those features which have strong collinearity into a single feature. Some of the feature extraction techniques that have been successfully used are Independent Component Analysis (ICA), Principal Component Analysis (PCA), Singular Vector Decomposition (SVD), Self-Organizing Maps etc. [46]. The difference between the features selection and the feature extraction method is that while the former allows us to select the most important features from the original feature set, later transforms the original features into a new feature set with a smaller number of features than the original feature set.

1.4.2. Normalization of Data to Remove Bias

When developing an ANIDS model, the sample data used for training may contain some values which are numerically very large as compared to other features. In such cases model can give the results which are biased towards these large values. Normalization of the numeric data is done to scale all numeric values between two fixed values. The scale used for the normalization can be [0,1[, [1,-1] etc. Normalization is used to avoid the bias in results due to large numeric values present in the dataset [48]. This is useful to prevent the incorrect prediction. Commonly used normalization methods include minimax algorithm and Z-normalization technique.

1.5. Performance Evaluation of Network Intrusion Detection Systems

To evaluate the performance of ANIDS, the parameters associated with confusion matrix are used [49]. There are various performance metrics that have been used to evaluate the ANIDS models which include :

(i) True Positives-True Positives (TPs) refers to the number of Intrusion alarms raised by ANIDS which turned out to be real intrusions. ANIDS should be able to detect maximum TPs or anomalies that are present in the network traffic to minimize the probability of intrusions.

(ii) True Negatives- True Negatives (TNs) are the negative alarms detected by ANIDS from network traffic. In other words, ANIDS treated these packets as normal and the prediction by ANIDS turned out to be correct. An efficient ANIDS should be able to detect maximum TNs from the network traffic. ANIDS generally do not raise any alarm for normal traffic but let it go.

(iii) False Positives- False positives (FPs) or False alarms(FAs) are the alarms raised by ANIDS as intrusion alarms. However, in reality there was no intrusion. Such kind of alarms are misleading and need human intervention. For an ANIDS to be called efficient, it should generate least or no FPs to prevent a human intervention. More FPs means more human intervention, which is undesirable in cloud environment. TPs and FPs constitute the alarm set raised by an ANIDS.

(iv) Accuracy - Accuracy is one of the most widely used performance metrics that is used to check the overall performance of ANIDS. Accuracy of a model is the ratio of number of correctly classified alarms (positive or negative) to the total alarms raised.

Accuracy of an ANIDS should be as close 100% as possible when expressed as percentage or as close to 1 when expressed as fraction.

(v) True Positives Rate-True positive rate (TPR) also known as recall or detection rate is the fraction of TPs detected out of all the true positives [50]. When expressed as fraction, TPR should be as close to 1 as possible. In terms of percentage, it should be as close to 100% as possible.

(vi) False Positive Rate- False Positive Rate (FPR) is the fraction of FPs detected out of all the TNs. FPs vary from 0 to 1 when expressed as fraction and 0 to 100% when expressed as percentage. For an ANIDS to be effective FPs should be as close to zero as possible to avoid any unnecessary human interaction in dismissing the alarms.

(vii) Precision- Precsion is the ratio of the total number of TPs detected by an ANIDS to the number of all TPs present in the dataset . Precision should be as close to 100% (1 for fraction) as possible.

(viii) F-Measure- F-Measure is the harmonious mean of the precision and recall. It is a statistical function for estimating the accuracy of a system by computing its precision and recall [51].

Table 1.5 shows the performance metrics for evaluating an ANIDS and the formulae for their calculation.

Table 1.5. Performance Metrics for evaluating ANIDS

Performance Metric	**Formula**	**Performance Metric**	**Formula**
Accuracy	$\frac{\text{True Alarms detected.}}{\text{Total Alarm}}$	True Negative Rate	$\frac{\text{Total T.Ns detected.}}{\text{Actual Positives}}$
False Positive Rate	$\frac{\text{FPs detected}}{\text{Atual Negatives}}$	False Negative Rate	$\frac{\text{FNs detected.}}{\text{Actual Positives}}$
Precision	$\frac{\text{TPs detected.}}{\text{All Positives}}$	Recall	$\frac{\text{TPs detected.}}{\text{All Positives}}$
		F-Measure	$\frac{2*(\text{Precision}*\text{Recall})}{(\text{Precision}+\text{Recall})}$

1.6. Challenges Faced by Anomaly Based Network Intrusion Detection System in Cloud Computing.

Cloud computing is enabled by the technologies like, virtualization, openness, and multi-tenancy. VMs are the building blocks of the cloud infrastructure which help in enterprise resource pooling. Openness allows the cloud resources to be accessed from anywhere in the world [35]. Multi-tenancy helps to host multiple organizational resources on different VMs but under the same physical host. All these enabling technologies also provide the security challenges as shown in the figure 1,4. However, there are some challenges that ANIDS faces during the detection process in cloud computing.

Since cloud hosts multiple tenants or organizational infrastructure, it is obvious that the incoming data from multiple organization into the cloud environment is huge. To detect network intrusion in cloud environment we need to have ANIDS that can detect at a faster speed and with greater accuracy. In case ANIDS is not able to keep up with the speed of incoming network traffic, ANIDS will get exhausted and drop some of the packets [52]. If these dropped packets turn out to be malicious packets, it can impact the cloud infrastructure and the data that is stored in it may get compromised. Thus, ANIDS needs to make decisions quickly and raise an alarm when a malicious packet is detected.

Accuracy is one of the parameters on which ANIDS is evaluated. Accuracy refers to the correct classification of the network traffic. Although ANIDS can detect the zero-day attacks, but the accuracy of ANIDS is still less and needs to be improved. ANIDS also results in more false alarms that require human intervention. In case of cloud computing where multiple tenants are hosted with thousands of servers in each tenant, it is necessary that human intervention is minimized. This is possible only if false alarms are minimized, if not eliminated completely.

One of the major challenges faced by ANIDS is the time complexity of the detection algorithm. Since the network traffic in cloud is huge and is high dimensional in nature, ANIDS must inspect and analyze all the features of a packet to classify the network packet as normal or anomaly. Time complexity can be reduced if ANIDS can make accurate decisions by inspecting fewer packets. This can be achieved by using dimensional reduction techniques.

1.7. Motivation of the Research

Security was a concern in early days of computing when computers used to work in isolation. Someone could break into the computer room and do any kind of malicious activity. This threat increased manyfold when computers started to communicate with each other. Today, when the information technology has no boundaries, any malicious activity can have impact on large numbers of servers spread across multiple geographical locations. This is especially true in case of cloud computing environment which hosts critical data and has no geographical boundaries [53].

Since cloud has multiple end points and is open in nature, it is vulnerable to cyber-attacks which can include DOS, DDOS, Sniffing, Injection attacks etc. This puts the critical data which is stored on the cloud servers at risk. Hence, it is extremely important that a CSP can detect or prevent the attacks in a timely manner to maintain the confidentiality, integrity and availability of data and the system

ANIDS are often used to detect the intrusion attempts in network including the cloud computing. However, given the fact that incoming network traffic of cloud is much more than that of a single network, ANIDS should be able to detect the intrusions in a timely manner and raise an alarm if any malicious activity is found. Moreover, given the types and number of new attacks that are increasing day by day, ANIDS should be able to detect the new attacks with greater accuracy. Hence, accuracy, false alarms and number of features are critical parameters for ANIDS in cloud.

In our study, we explored various techniques that were used in developing the ANIDS in cloud environment [83-91] and evaluated their performance on various performance metrics like accuracy, false alarms and the number of features used to make predictions. We identified that there was some scope to increase the accuracy of ANIDS models for cloud and at the same time reduce the number of features to get the desired accuracy. This is an important requirement for ANIDS in cloud computing. While increasing the accuracy, we also needed to reduce the false alarms for ANIDS in cloud computing.

Motivated by the fact that security concerns in the cloud environment need to be addressed by having an ANIDS which can give better results than the models that have been proposed earlier, our aim was to develop ANIDS using Supervised MLTs which makes predictions using fewer features and with better classification accuracy.

1.8. Research Problem

Motivation to make the cloud a more secure place for the user data lead to the formation of a more specific research problem. This was done by identifying the techniques that were used to develop the ANIDS for cloud and analyzing their performance.

Due to the multi-tenant nature of the cloud, there is a high volume of incoming network traffic that enters the cloud. This traffic is high dimensional in nature and needs to be processed by an ANIDS to detect anomalies. ANIDS must process, inspect, and analyze all the feature of each network packet to make the decision. The high volume, high speed and high dimensionality of the traffic can cause the ANIDS to exhaust and drop some packets which can cause intrusion packets to enter the cloud. If a single such malicious packet enters the cloud environment, it can impact not only a single location but across multiple geographical locations. Thus, an efficient ANIDS needed to be developed to detect such intrusion attempts and to prevent or at least minimize the damages by these intrusions.

Researchers have used various features selection and feature extraction techniques to reduce the number of features which can minimize the overhead on ANIDS [39]. The problem of high dimensionality is more critical in cloud computing as compared to the on-premises network. High dimensionality of the network traffic also increases the time complexity of the detection algorithm, which can cause delay in detecting the intrusions. Hence, we performed the literature survey to identify the dimensional reduction techniques that were used and how the features were optimized to get better accuracy. We identified some gaps in the research and converted them into the research problem. We observed that less effort was made to optimize both the accuracy as well as the number of features. We treated this problem as a multi-objective optimization problem and proposed a new optimized model for ANIDS.

Hence, the problem that this study focused on was to reduce the feature space of the network data and at the same time increase the accuracy of supervised learning based ANIDS. This could reduce the time complexity of detection and greatly improve the performance of ANIDS for cloud computing.

1.9. Research Objectives

After identifying the research gaps, the research question was "how to develop an ANIDS based on supervised MLT that can make accurate prediction using fewer features set". This research problem was divided into four objectives.

RO1: To study the current intrusion detection techniques and how they are implemented in the cloud.

RO2: To identify and analyze recent intrusion detection techniques in cloud environment with different parameters of interest.

RO3: To develop and propose a technique for network intrusion detection to collaborate with each other for effective intrusion detection.

RO4: To evaluate the performance of the proposed collaborative network intrusion detection technique and compare with other existing techniques.

1.10. Research Questions

Research question is a specific issue or a problem that needs to be addressed. Each of the research objective was further broken down into the research questions to be addressed. A research objective could need one or more than research question to be addressed. The mapping of research objectives to research question is shown in figure 1.12.

RQ1: This research question was addressed during the literature review. The specific problem that this question addressed was to identify the broad range of techniques that were being used in developing the intrusion detection techniques. This includes MLT and statistical techniques. This question also addressed the identification of dimensional reduction techniques as well as optimization techniques and their impact of the accuracy of classification. We also identified various data sets that are were used by the researchers in developing and evaluating the NIDS models and the parameters used for evaluating the models.

RQ2: This research question was framed to identify the methods that were used to evaluate the performance of a NIDS. The parameters of interest as well as their desired values were identified. Methodology used to address this question was literature review.

RQ3: This research question addressed the problem of identifying the techniques specific to cloud environment. The challenges for a NIDS in cloud environment were identified and various methods proposed by researchers were studied. We also identified the level of performance that NIDS in cloud computing that has been achieved using MLTs. This served as a baseline for our research work. Methodology that was used in addressing the RQ1 and RQ2 was the literature review. We selected papers published during last more than 12 years and reviewed as well as synthesized those papers as well as prepared the literature matrix.

RQ4: This research question was mainly focused on identifying the feature selection methods that could give the best accuracy on a dataset. Since, we wanted the feature selection method to be classifier independent, we chose statistical based feature selection method. We also identified feature extraction method that performed best of the dataset. The methodology that was used during phase was comparative analysis. We selected and compared the multiple feature selection methods and performed the comparative analysis of their performance to arrive at a conclusion. Same methodology was adopted to select the best feature extraction technique.

RQ5: This research question was addressed in the proposed model. The main question was to keep the feature set minimum but increase the accuracy of the model. This was done by treating the problem as a multi-objective optimization problem and using optimization methods to solve it. Methodology used to address this research question was the proposed work which was based on optimization methods and yielded better accuracy.

RQ6: Since the proposed model was a multilayer model, the formula for parameters that were identified in RQ2 needed to be modified for this model. This was done during the development of the proposed model

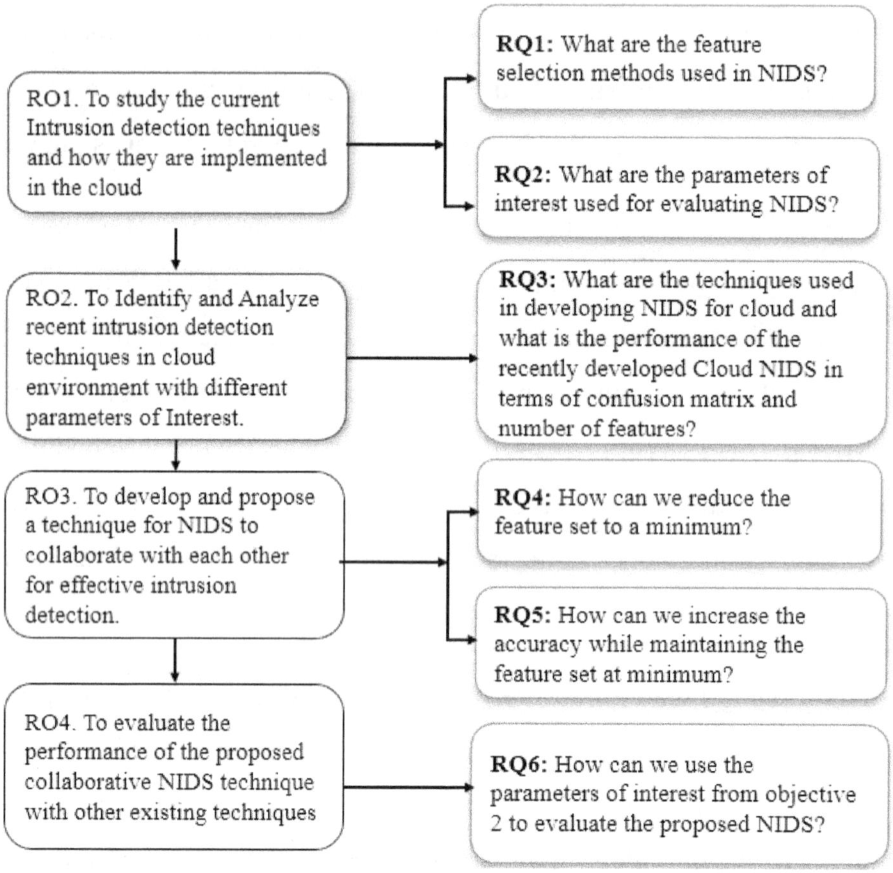

Fig.1.12. Research Objective to Research Question Mapping

1.12. Thesis Contribution

While the main contribution of the thesis was that we proposed a model that performed better in terms of accuracy, FAs and number of features, we believe some of the features that are worth highlighting include:

a) Minimized the Feature Space by using Optimization Method

To select the most relevant features, we used the optimized correlation-based feature selection method. However, we went a step further to use the feature selection followed by feature extraction method to transform the selected feature set into a new and reduced feature set. This was done to remove the collinearity between the independent variable to get more accuracy. To optimize the process of search in feature space, we used Particle Swam Optimization algorithm (PSO). We also used SVD for feature extraction which gave better results than PCA and LCA. We used genetic algorithm to optimize SVD dimensions as well as the hyperparameters of classification algorithms. We were able to reduce the feature set to a minimum of 7 and at the same time achieved better results in terms of confusion matrix as compared to similar models that were proposed in the past.

b) Optimized Machine Learning Models

To increase the accuracy of the proposed model, we used multiple MLTs to create multiple layer model, where normal traffic and anomalies were further classified to increase the classification accuracy. In proposed model, the processed network data from layer-1 was classified as normal and anomaly. Normal and anomaly packets were then sent to layer-2 classifiers to further classify normal and anomaly traffic. To get the improved results, we first optimized the parameters of these MLTs individually using the genetic algorithm. Once these MLTs were optimized, they were integrated to create proposed model. Model was then trained and tested using NSL-KDD dataset.

c) Improved Accuracy with Least Feature Space

We used the confusion matrix along with the number of features to evaluate the performance of the proposed model. Our optimized model resulted in improved performance as compared to the earlier models that were proposed for cloud computing. This was mainly because we used multiple layers for classification. The re-classification of normal and anomaly traffic by layer-2 model resulted in the improved performance of our model. The main contribution was achieving better accuracy with fewer feature set. Hence, our proposed model is suitable for cloud computing.

1.13. Structure of the thesis

Thesis was divided into multiple chapters, where each chapter focused on a specific outcome Following is the summary of these chapters.

Chapter 1: Introduction

In this chapter we have introduced the concept of cloud computing, its features as well as architecture of the cloud. This chapter introduces the concept of network intrusion detection systems and its need for cloud computing. The chapter also discusses in detail the types of Intrusion detection system and their limitations. Further, the importance of machine learning technique along with the use cases of feature engineering and normalization is discussed in brief. The performance metrics that are used to evaluate the performance of ANIDS is also discussed in brief. This chapter also covers some of challenges that ANIDS faces in cloud computing environment. This chapter concludes with the research objectives and thesis contribution.

Chapter 2: Literature Review

In this chapter we have discussed and synthesized the existing literature published since last twelve years in the field of cloud environment. This literature was classified based on type of implementation of intrusion detection system. In this chapter, we have also presented the literature review matrix along with the gaps in each research. We have also listed the research papers that were further studied and formed the basis of our work.

Chapter 3: Performance Evaluation of Feature Selection Methods

In this chapter we have discussed the concept of feature engineering techniques that are used to reduce the dimensions of the dataset. In this chapter, we have done the comparative analysis of four correlation-based feature selection methods. This chapter has also presented the detailed analysis of the performance of these four feature selection methods. The chapter ends with the concluding remarks.

Chapter 4: A Comparative Study of Feature Extraction Techniques for Network Intrusion Detection System in Cloud Computing

In this chapter we have discussed the feature extraction methods. In this chapter we have performed the comparative analysis of PCA and SVD and ICA to identify the feature extraction technique with better performance.

Chapter 5: Recent Approaches used in Anomaly Based Network Intrusion Detection Systems

In this chapter we have discussed in brief the various machine learning techniques in detail that were used to develop the NIDS is recent years. We have also analyzed their performance as their performance as documented in the literature. The objective was to identify the MLT that yielded better accuracy.

Chapter 6: Proposed Collaborative Bases Network Intrusion Detection System in Cloud Computing

This chapter has discussed the methodology used in the development of proposed model. The details of the experiment used in developing the proposed model are also discussed in detail. Finally, the performance of the proposed model is analyzed and compared to the earlier models to arrive at the conclusion.

Chapter 7: Conclusion and Future Work

This chapter has provided the concluding remarks and the scope of the future work in cloud security.

Chapter 2

CHAPT ER 2: LITERATURE REVIEW

2.1. Introduction

Literature review is one of the most important phases of the research. During our literature review we reviewed and synthesized more than 150 research paper from last 12 years and identified broad research topic to be based on the security solutions for cloud environment. After synthesizing the papers, we identified the techniques that were used to develop the solution for cloud security. Next was to identify the gaps in the literature and the improvements that could be made to the present work. So, we focused on the topic of NIDS and identified some of the gaps as well as the improvement areas. We narrowed down the topic to a specific problem statement. More focus was given to the papers which provided solutions for cloud environment. To solve the research problem, it was divided into four research objectives. Each of the objectives had one or more than one research question to address to complete that research question.

The literature was organized based on the type of proposed implementations and literature matrix was prepared based on the outcome of literature review. Since our research was focused on the cloud computing, we identified the papers that provided solutions for cloud security and could form the basis of comparison of our work. We prepared a separate list for those papers and analyzed them further. Figure 2.1 shows the methodology that was used during literature review.

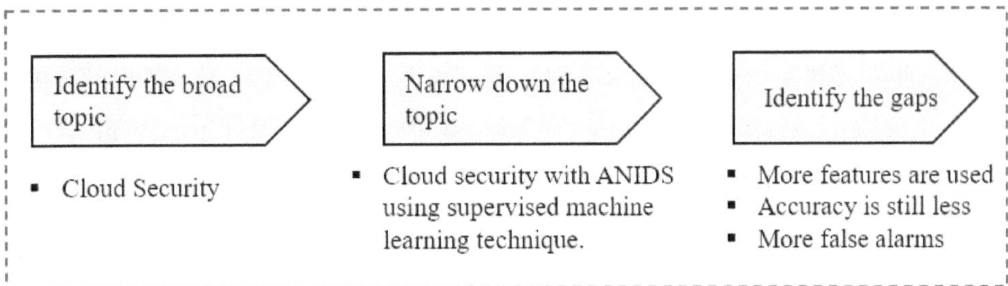

Fig.2.1. Literature Review Process Used

2.2. Hybrid Intrusion Detection System

Neither SNIDS nor ANIDS alone can detect all the threats in an IT environment. However, if we integrate these two techniques, we can improve the performance of NIDS. In this section we have discussed the literature related to hybrid NIDS. We define the hybrid technique as the one that combines both, the misuse detection method and anomaly detection method to enhance the detection performance. If implemented properly, hybrid NIDS should be able to reduce the probability of cyber-attacks. In most of the cases misuse detection method has been used first to detect the known attacks and then anomaly detection method has been used to detect unknown attacks. This way we can filter the known attacks first and reduce the time of anomaly detection. Snort is one of the common SNIDS techniques which is used for detecting the known attacks. However, Snort can be customized, and modules can be integrated with Snort to enhance its functionality. Some researchers have integrated anomaly detection modules with Snort to enable it to detect anomalies also [54].

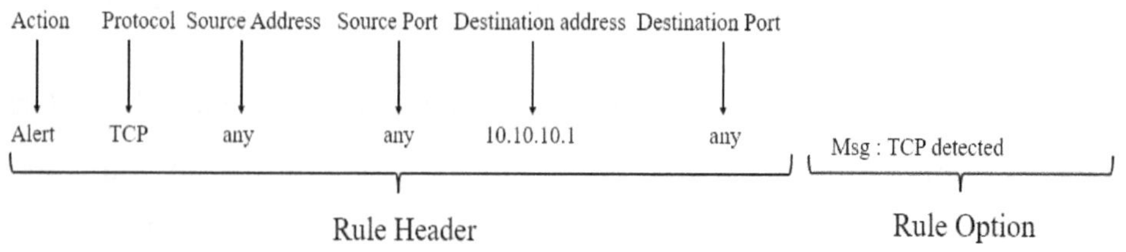

Fig 2.2. Snort Rule Structure

Snort is a signature base NIDS that can be used to identity the known attacks based on various parameters like, source IP, destination IP, port etc. However, it will not be able to detect any zero-day attack or a variation of already known attack. Snort configuration file contains rules to identify anomalies. Snort can be run in three modes, i.e Sniffer mode, Packet logger mode and NIDS mode [55]. Sniffer mode allows TCP packet headers to be directed on the screen. One can observe any anomaly in the packet headers. Packet logger mode allows the output of the commands to be directed to a file located on the local disk for later analysis. To configure Snort to inspect all the packets on the network, Snort needs to be configured for the NIDS mode. ***Snort.conf*** is the main file where the rules for a packet to be considered as an anomaly are defined. Every packet on the network is compared with the rules in the snort file. If a match is found, the packet is treated as an intrusion and alarm is raised. Snort rules contain two parts i.e Rule header, and Rule option. Rule header contains fields like Action, Protocol, Source IP address etc. Rule option define what message needs to be displayed in case any rule is matched. A simple format of the rule is shown in the figure 2.2. During our literature review, Snort was found to be most used misuse method along with other anomaly techniques [55].

Modi et al. [56] has used Snort and Apriori algorithm to detect known attacks as well as variance of known attacks. The signature databases was used to detect the known attacks, Apriori algorithm was used to generate the full signature from a partial known signature and update the Snort rules. If the network traffic is high, it has been shown that the Snort drops packets which could increase the risk of intrusion. Above all, this model will also not be able to detect completely unknown attacks where even partial signature is not known. Updating Snort database with new signatures is also a challenge. Apriori algorithm was first proposed by Agrawal and Srikant [57] in 1994. This algorithm uses the prior knowledge to find the frequent itemset in a dataset. It uses two processes to perform the task. First it generates the candidate item sets and calculates the support count for each of them. The items that have the support count less than the minimum support count are dropped. This is done repeatedly until frequent sets cannot be generated.

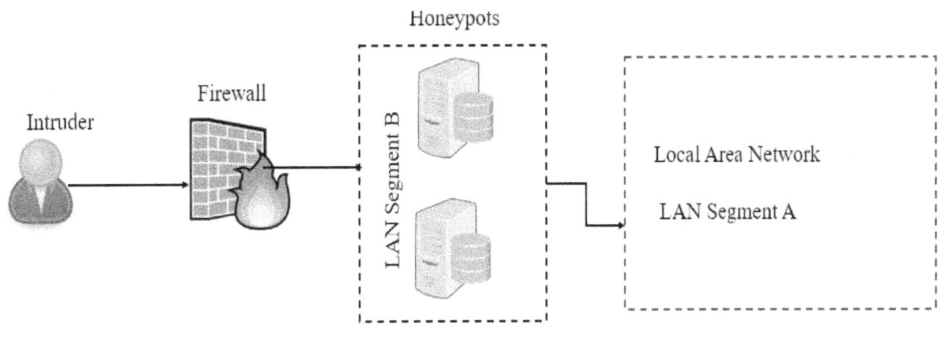

Fig.2.3. Network with two Honeypots

Honeypots play an important role in trapping the intruders. They can be used to keep the intruder engaged and understand the activities of an intruder which can help to design the rules for a NIDS.

A honeypot is a deception technology in which the device is presented to the intruder as a normal machine with useful information. It is used to lure the attacker to launch an attack on it. Honeypot does not take any action or prevent the attack. it is a dummy machine which is used to gather the information about the intention and the behavior of the intruder. From all this information, we can know the capacity of the intruder and the specific intentions behind the intrusion attempt. Low interaction honeypots gather less information than high interaction honeypots. Low interaction honeypots are simple to implement as they have limited level of emulation for the attacker activity. However, high level honeypots gather more extensive information about the attacker as they interact with other systems and applications [58]. Figure 2.3 shows the placement of two honeypots behind a firewall. Honeypots are normally isolated from local area network and kept in a separate network segment than local area network [47]. An organization can have one or more than one type of honeypots with sensors to log the behavior of intruder.

Rizvi et al. [59] have proposed combination of the technologies like NIDS, HIDS, honeypots and firewall to detect misuse and anomaly based attacks. High interaction honeypots were used to detect known attacks and low interaction honeypots were used for unknown attacks. Proposed model used genetic algorithm to analyze the unknown attacks and fuzzy logic to calculate degree of truth. The architecture also maintained a database of detected attacks for future use. However, this model does not present the evidence for the performance which could have been useful. Also having multiple IDS (Firewall, SIDS, NIDS, honeypots) can make the IDS model more complex and at the same time result in delay in processing as well as delivering of network packet

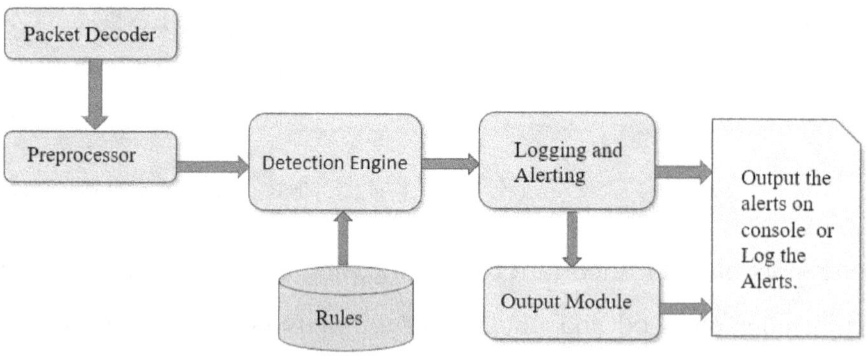

Fig.2.4. Architecture of Snort

Snort has five major components whch include Packet Decoder, Preprocessor, Detection Engine, Alerting and logging system and Output module, The architecture of Snort is shown in figure 2.4. Packet decoder module receives the packets from the network and does the initial analysis of the pasket. Preprocessor module of Snort performs the packet defragmentation, TCP flow reasembly and so on [23]. The main component of Snort which performs the intrusion detection function is Detection Engine which is guided by the rules in the snort.conf file. However, it is possible for programmers to enhance the functionality of Snort to plugin the customized modules in preprocessor. For example, an anomaly detection module can be integrated with snort. **Gómez et al.** [60] has proposed the NIDS model in which they used a hybrid method that could increase the performance of Snort. In this work anomaly detection preprocessor was integrated with Snort to detect known and unknown attacks. It used MYSQL at the backend to store the normal data during training period which was then converted into the profile using moving average method. MYQL also stored anomalies during the detection phase. During detection phase Snort used MYSQL database to detect anomalies. Test was performed using DARPA network and reduction in false positives was observed. However, deciding on the time frame for the moving average could be a challenge. With different time frames we can have differenr moving averages.

When the network traffic or IT infrastructure is dynamic in nature, NIDS needs to adapt to these changes and perform consistently and with high precision under the changing environment. This is especially true for the cloud computing where traffic is dynamic. **Rajendran et al.** [61] advocated the property of adaptability, scalabilty, dymanic nature and efficiency of IDS. They propsed an NIDS model which was implemented with dotnet at the front end and SQL at the backend. They claimed that the scalability and self adaptivity can be achieved by deploying IDS at network and also at every host in the netowork. However, deploying IDS in network and at every host is clearly an overhead on the network. Also the network packets will be be processed multiple times by IDS which could cause delay in detection of possible intrusions.

Some intruders also exploit the weakness in the protocol stack to launch the attacks like Sync flood, fragmented packet attack etc. [62]. In such a case modeling the behavior on the basis on traffic only will not help. We need to have a NIDS which can monitor and analyze each network packet individually and find out the abnormalities in the packets. Packet Header Anomaly Detector (PHAD) was the first preprocessor that models the pattern on the basis of protocols rather than the user behavior [63]. Thus, it can detect those intrusion that take place due to the weakness in the protocols. It is a time based modeling and assumes the traffic is dynamic and changes quickly. Hence, it is more adaptable in cloud computing. PHAD also raises a single alarm for a similar intrusion, thus reducing the number of of false alarms which is also an advantage in cloud computing. PHAD analyses every packet that passes through the network and does not distinguish between incoming and outgoing packets. It models 33 attributes of the packet header like TCP header, IP header, ICMP header etc. to detect the anomalies. Like PHAD, NTAD (Network Traffic Anomaly Detecor) is another preprocessor which can be used with Snort. It also models the packets similar to snort but works in two phases. In first phase, it uses filtering to distinguish the sessions and in the second phase it does the modeling. The modeling phase uses the protocols like IP, TCP, FTP, SMTP etc. PHAD and NETAD can be integrated with snort to enhancce its detection ability . **Aydın et al.** [63] futher contributed towards the enhancemnet of Snort to detect misuse as well as anomaly based intrusions. It integrated two pre-proccessors PHAD (Packet Header Anomaly Detector) and NTAD (Network Traffic Anomaly Detector) with Snort. They model the beviour on the basis of protocol by looking into the packet header and assume quick changes in network traffic. Looking at each packet can surely detect more attacks as compared to only looking at the overall traffic. However, the drawback of this method is that NIDS needs to examine each packet individually which could cause Snort to drop the packets and also is an overhead. This is especially true when the volume of incoming network traffic is high and dynamic in nature as in case if cloud computing.

In cloud environment where the incoming traffic is huge, Snort may not be able to keep up with the incoming network traffic and drop some of the packets. The basic reason being that Snort contains thousands of rules and every packet needs to be compared with every rule. Hence, with every new rule, the time complexity of algorithm also increases.

Uddin et al. [64] have proposed distributed intrusion detection technique by distributing a subset of rules (frequent attacks) among multiple Small Signature Databases (SSD). A complementary database stored all signatures identified during training period. If an attack became more frequent it was taken from complementary database and added to most frequent database by mobile agents or it was removed in case it became less frequent. This approach is useful in reducing the Snort database size and speeding up the detecting process. However, deciding on what constitutes a "frequent" attack and how long it to be considered a "frequent" is something to be considered. However, every packet needs to be examined by multiple Snorts which could cause delay. Another challenge is that the communication between the complementary database and SSD needs to be secure to avoid any men in the middle attack. If any of the SSD is not timely updated or is down, it can result in incorrect predictions as SSD will not get updated.

Misuse as well as anomaly detection methods have their pros and cons and. While misuse methods can be good at detecting known attacks, anomaly can be good at detecting unknown attacks. However, if these two algorithms are combined, we can get better results as compared to their individual implementation. However, it is useful to use misuse method before anomaly method to detect known attacks prior to unknowns, **Kim et al.** [65] proposed the hybrid approach to combine these two methods where the normal output of the misuse detection was fed to the One Class Support Vector Machine (1-SVM). It used DT to train the misuse detection model given the known attacks. DT was created by repeatedly dividing the data into separate data sets until a stage was reached where the leaf node belonged to a class. Each normal subset of the traffic from the DT was used to train the multiple 1-class SVM and to detect intrusions. If any attack was detected at the DT level, it was flagged as an intrusion there only. For a high featured data, DT will not be an appropriate method for misuse detection. If there are already clusters in the data, DT will not consider it and segment the data which can increase the computational cost. DT can also become complicated in case of high dimensional data. However, DT is a good candidate for using parallel processing technology to reduce time.

Chiba et al. [66] proposed an NIDS model that used Snort as the NIDS and and an

Optimized Backpropagation Neural Network (OBNN) to detect unknown attacks. It also used a central database to store all the detected attacks from NIDS. Here, NIDS was trained using normal and anomaly data. The central databse was used by Snort and anomaly detector to detect intrusions in future. In this case, frequent updation of the Snort database with new attack signatures is going to be a challemge.

Two properties that are important for any NIDS are its accuracy and consistency. Accuracy means low false positives and negatives. Consistency means that the NIDS should give the accurate results irrespective of number of observations. These two features are also important for a NIDS to protect the critical in cloud computing.
Jabez & Muthukumar [67] proposed Neighbor Outlier Factor (NOF) to increase the accuracy of NIDS. This model was trained using large dataset from distributed environment that increased the precision of the model. It used combination of Snort and k-Nearest Neighbor method (k-NN) to detect the outliers and classify them as intruders. However, k-NN can be computationally very expensive when the training data is large, and it can result in the late predictions which could affect the timely detection.

2.3. Agent Based Intrusion Detection System

In this method, mobile agents (software modules) are used by the NIDS to perform certain tasks. The advantage of such technique is automation. Multiple agents can be assigned multiple tasks which could speed up the detection process. However, one of the disadvantages of such an approach is that the communication cost is increased as there needs to be coordination and communication between the agents. Another challenge is to make sure that the agent has performed the assigned task. In some cases, agents may themselves get attached by malicious users.

An intrusion detection system needs to adapt to the dynamic network environment and needs to respond quickly when an intrusion is detected. This is possible when multiple modules are implemented to perform the multiple tasks in parallel. This can speed up the intrusion detection process and raise an alarm in a timely manner to take the action.
Teng et al. [68] proposed NIDS model in which multiple agents were used to perform various tasks, Event generator identified suspicious events and Event detector classified

them based on protocol like, TCP, UDP, ICMP. DT and SVM were used as classifiers. Multiple agents performing the same task were grouped together. Having multiple agents performing the same task increased the detection and response time. However, as the numbers of agents increased, the coordination between the agents would be a challenge and may cause a communication overhead. This would also make the model complicated. Agents also need to be monitored for their performance and any failure.

One of the areas that has been of interest to the researchers is Human Immune System (HIS). HIS can be thought of as an intrusion detection and prevention system where the body detects the abnormalities and generates the antibodies to fight the foreign bacteria. Artificial Immune System (AIS) is inspired by HIS and tries to mimic HIS to detect and fight the intrusion [69]. Various AIS algorithms like Clonal Selection Algorithms and Immune Network have been proposed and successfully used to detect the intrusions.

Zhang et al. [70] proposed the hybrid system which integrated AIS algorithm and misuse detection technique. This model was based on antigens, antibodies, and vaccination. Misuse detection technique was used to detect known attacks. During the anomaly detection process, the antibodies were created which became mature and improved the detection process. However, with the increased antigen set for training the space complexity can be a concern.

Afzali Seresht & Azmi [71] proposed another agent-based NIDS which also used the concept of HIS. In this model multiple agents were used to perform the task. However, what made it more interesting was that the concept of polymorphism was applied at the agent level. Agents were evaluated for their detection performance and the agents that performed best shared their knowledge (cloned) with other agents. To keep the population of agents under control, weak performing agents were killed. However, the increased number of agents made the system more complex. Another drawback is that we also need to add another module to evaluate the agents which could also consume computational time as well as memory and ultimately delay the detection process.

Teng et al. [72] proposed a multi-agent collaborative NIDS for anomaly detection. However, in this framework, we had an agent (Statistical detector), which alarmed an anomaly if the number of packets received in a time interval was more than a pre-defined value. Fusion center were another group of agents that accepted anomalies from different agents and correlated them. Given the nature of the cloud where the volume of the traffic is not constant and can have sudden bursts in the traffic at different intervals, this agent will raise frequent false alarms which may need human intervention.

When using mobile agents in NIDS, two issues may arise. First, how can we make sure that the agent was not compromised and second, how can we make sure that the agent performed the task that was assigned to it. This means that the mobile agents themselves need to be monitored by some software module which needs to keep track of all agents. **Idrissi et al.** [73] proposed a mobile agent-based NIDS system. In this system, the mobile agents contained the list of IP address of the servers that they needed to contact to identify if they are malicious or not. Mobile agents also left traces (Signature) of execution on servers which were created using the public key of CSP and used by agent during execution on servers. This signature may consist of information like identity of agent arrival time, execution time etc. This information made it possible for the CSP to verify the execution of the agent on the server. After all the information from all the servers was collected, it was sent to CSP. This method can be useful to make sure that the agents have executed. Since it used the public and secret key combination, it can be secure. This model used a single mobile agent which travels multiple cloud servers, which would cause delay if list of severs is large. Instead, we could use multiple agents with subset of IP addresses for each of them and do the task in parallel. Multiple agents can increase detection performance of NIDS. The concept of automatic computing can also be used which involves the self-healing, i.e. in case of any failure, the agent will try to correct itself using the knowledge that is available with it at that point of time.

Firewall is the most used security device to protect the unauthorized traffic from entering the network. Firewall is a combination of hardware and software which is deployed at the perimeter of the network. Firewall maintain a set of rules against which the incoming traffic is matched [74]. The rules are based on source IP address, destination IP address, domain name etc. If any of the incoming packets matches the rules in the firewall, an action associated with that rule is activated. One of the drawbacks of firewall is that it cannot detect the insider attacks as it is installed at the perimeter. Firewall also requires expert knowledge to configure, and it cannot detect zero-day attacks [36].

Toumi et al. [75] proposed agents, Jade container, Snort, Firewall and SQL database to implement this technique in NIDS. If any intrusion was detected by Snort it was logged into local database which was read by the static agent. Cloned agents collected this information and migrated to Virtual Machine Monitor (VMM). These were used to generate new rules by Jade main container and applied through firewall. While this model uses the automatic computing to apply rules, it will not be able to detect the new attacks as it uses Snort only. This work also does not show the experimental results which could have been useful in assessing the effectiveness of this solution.

2.4. Collaborative Intrusion Detection Systems for Federated Clouds

The main aim of the federated NIDS is to detect the intrusions in large networks like multiple clouds and avoid generation of false alarm. In these systems, multiple NIDS generate the alems which are then correlated to arrive at the decision. This helps when different networks have NIDS from different vendors which are capable of detecting different attacks. These techniques use ensemble based algorithms to fuse the attack information between two or more NIDS [76]. NIDS from differen vendors help in detecting the different types of attacks as well as zero day attacks. A single NIDS may not be able to detect all the attacks in the cloud. The federated clouds are a group of clouds that have established trust and are able to share the cloud resources with each other. The resources may include data, VMs, and the alarms generated by the NIDS.

When there are multiple decision (beliefs) from multiple federated cloud NIDS about the intrusion detection, NIDS in a cloud environment needs to correlate these multiple beliefs from the multiple cloud NIDS to reach at a single decision. Dempster-Shafer is one of the algorithm that combines multiple beliefs to reach at a single belief [77]. Dempster-Shafer uses prbabilitic theorey and calculates the degree of trust to arrive at a decision.

MacDermott et al. [78] has proposed the federated NIDS model where the clouds can share the attack information. If a cloud domain was not able to detect the attack, the special node called cloud broker was used to consult the trusted domain to see if it had any information about the attack and used Dempster-Shaefer theory to fuse the information together and make the decision. The main issue with this model is that the communication between the two clouds need to be secure which can be achieved through encryption. However, encryption could cause delay in detection and response [79]. Also, men in the middle attack can be another pronlem with this kind of model.

The main technology that enables the sharing of the resources in the cloud environment is the virtualization. Using virtualization, multiple VMs are created under the same physical host. VMM or hypervisor is responsible for creating an abstraction layer above the hardware layer [20]. VMs are an important component of cloud. However, VMs also increase the surface area of attack as multiple VMs exist together under a single host. Deploying NIDS on VMs can make NIDS vulnerable to the attack if attack is carried out at hypervisor level. Such an attack can impact multiple VMs at the same time.
Huh [80] proposed a collaborative NIDS model with federated clouds. The proposed architecture used Elementary detectors (EDs) which were used to detect anomalies and store in local database. These EDs were installed on a separate domain and not on VM to be monitored. It used Virtual Machine Introspection technology to fetch the information from each VM. Initially the raw alerts generated were processed by dispatcher to find any pattern within a domain. This architecture also had a NIDS manager which correlated all the alerts from all cloud domains and sent attack information to each domain to remain updated. All the communication happened in encrypted manner to ensure confidentiality. Clearly, this framework could make the NIDS more secure. Since ED needs to communicate over the network with VMs, any delay in collection and analyzing the logs can result in delay in decision making. Even for a false alarm, the data will be sent three hops upstream which is a wastage of bandwidth. If the false alarms increase, it is going to generate the unnecessary traffic which can impact the network performance for business-critical applications.

When using multiple NIDS in a collaborative manner, there needs to be a method to check the reliability of the partner NIDS. This is necessary to see if they can provide the reliable information about the attacks. Also, in case the partner NIDS are not able to provide feedback in timely manner or if the partner NIDS has incomplete information, it can have an impact on aggregation process and can result in incorrect NIDS decisions. **Fung & Zhu** [81] made an effort to address these two issues. The proposed model had multiple NIDS, where each NIDS maintained a list of neighboring NIDS that it needed to communicate with. Each NIDS sent a test message to its neighbors and if the answer to the test message matched the expected result, it was put in the "Consultation" list else it was removed from this list. During the detection process, the network packet was first matched to local signature database, if no match was found then NIDS communicated with consulting NIDS to make the decision. While this model addresses the NIDS trust issues, the "test" messages can create communication overhead. As anomaly methods creates lot of false positives, for each such false alert, it will have to communicate with neighboring NIDS. This can result in unnecessary communication between multiple NIDS.

Ficco et al. [82] proposed the distributed NIDS to detect large scale attacks. This model was implemented at three layers, customer layer, cloud layer and federation layer. Each layer had three sets of agents i.e. probes which gathered the data and detect anomalies, agents which received the data from probes, normalized it and sent it to security engine for correlation. If security engine at customer layer was not able to make decision, it sent data to security engine at cloud layer to get any feedback. Finally, if cloud security engine was also not able to make decision, it sent data to federated cloud. Multiple security engines were used to improve performance. Having three layers could increase the knowledge of NIDS but at the same time, it will also cause delay and communication overhead in case of increased false alarms.

2.5. Intrusion Detection System for a Single Cloud

Idhammad et al. [83] has proposed a distributed NIDS model for cloud computing. Network traffic was collected at edge servers at regular intervals and stored in storage server. Preprocessing was done on the stored data using minimax algorithm. NBA was used to classify this data as anomaly or normal. Anomaly data was then synchronized with the centralized server where random forest algorithm was used to classify the anomaly data into various types of attacks. Test on the data yielded an accuracy of 97%

using 14 features. Since the intrusion detection system was placed at the edge server, this NIDS won't be able to detect the internal attacks. Also, data synchronization from edge server to the central server will consume a lot of bandwidth and needs to be secured.

With exponential increase in the cloud network traffic, there is a need to have NIDS that can process this high dimensional network traffic quickly and accurately. For this reason, deep learning techniques have been proposed by various researchers, which have yielded good detection results as deep learning can mine a large volume of data. **Abusitta et al.** [84] proposed autoencoders to reduce the dimensionality of data and increase the accuracy. Logistic regression-based predictor was applied at the last hidden layer of the autoencoder to classify the data. This work claimed to have achieved an accuracy of 92.5%. However, the number of features selected for classification is not mentioned. Training the model that uses autoencoders is computationally expensive. Also, there is still a need to improve the accuracy of this model.

Zhang at al. [85] proposed a Many Objective Optimization Technique method to optimize accuracy, recall, precision, false alarms and number of features. Information gain was used to rank the features in NSL-KDD dataset and the subset of features above a certain threshold were selected. The selected features were then evaluated using k-NN based wrapper method. Finally, 30 features out of 41 were selected for the classification. Test results show that an accuracy of 93% was achieved on SVM and 84% was achieved on Naïve Bayes algorithm. However, Inspite of using multiple feature selection algorithms, number of features is too high, and accuracy is low.

Samriya et al. [86] proposed a model to avoid the issue of overfitting and reduce the computational cost. It used feature selection method to reduce the dimensionality of dataset. They used Fuzzy Cluster Means (FCM) and Spider Monkey Optimization Algorithm (SMO) for feature selection. FCM was used to create the cluster of instances based on Euclidean distance and SMO algorithm was used to optimize the search of the feature space. Classification was done on multiple classifiers and the highest accuracy of 86% was achieved on ANN. However, the decision to decide on the number of clusters which represent the number of features in this case is also a challenge.

Rabbani et al. [87] has proposed NIDS for cloud computing based on the user behavior. During the feature engineering process, PCA was used for dimensionality reduction. Non-numeric features were transformed to numeric type and minmax algorithm was then used to scale the data. PSO algorithm was used to optimize the Probabilistic Neural Network (PNN). PNN was trained using the training dataset and testing of the model yielded an accuracy of 96.4%. However, using multiple optimization algorithms can be computationally expensive and more complex in this case.

Krishnaveni et al. [88] has proposed NIDS for cloud computing based on ensemble feature selection method and ensemble classifier. Five feature selection methods were used and the union of the optimal feature set from all these methods was used as the final feature set for classification. Ensemble technique using four different base classifiers and voting method as metaclassifier was used for final classification. NSL-KDD dataset was used for the experiment with a final feature set of 10. This model achieved an accuracy of 96%. This model was based on the fact that multiple machine learning algorithms when used in ANIDS increase the accuracy as compared to a single algorithm. However, having multiple classifiers makes the model complex and computationally expensive. Ensemble model will also increase the detection time due to multiple MLTs. Having more feature selection methods is going to increase the training and detection time.

Thilagam et al. [89] proposed NIDS for cloud environment using RNN. The parameters of RNN were optimized using Ant-Lion optimization technique. During the feature selection, 22 most relevant features out of 42 were selected for the classification. Using this modeled, an accuracy of 94% was achieved. However, RNN are slower as compared to other Neural Networks.

In cloud environment, huge volume of incoming and outgoing network traffic with large number of features demands that NIDS should process the data at a rate which is higher than the incoming data date. Reducing the number of features in data can achieve this objective as NIDS will have to process fewer features. Removing irrelevant features or those features which have less or no impact on the class variable can also help to achieve higher accuracy, detection rate and make NIDS model less complex [28]. **Mayuranathan et al.** [90] proposed NIDS based on deep learning. During preprocessing, heap optimization algorithm was used to clean the data. For feature selection, Chaotic Red Deer algorithm was used to select most relevant features. Deep learning method was used for the classification which produced an accuracy of 97%. However, the heap optimization takes too long and is computationally expensive. Such technique can be an overhead for the cloud environment.

Kaushik et al. [91] performed the evaluation of various feature selection techniques to evaluate the performance of NIDS on these features using ensemble classifiers. An accuracy of 96.29% was achieved using ensemble algorithm and Chi-square as the feature selection method.

In the cloud environment, large volume of data and high dimensionality can cause the problem of "Curse of dimensionality". To avoid this problem, we need to select and extract the most relevant features for classifying the data. The processing and memory requirements can be huge in case of large dimensionality. To make model more efficient, we can use feature extraction techniques to get a new but fewer number of features. **Mahmood & Agrawal** [92] proposed NIDS model that used PCA for feature extraction. Once the dimensionality of the data was reduced; it used backpropagation method for the classification of the data. Since the number of features were reduced, it solved the problem of overfitting as well as it decreased the detection time of NIDS. This can improve the overall performance of NIDS.

Table 2.1 shows the literature review matrix which highlights the methods that were used and the gaps that were found. Table 2.2 shows the challenges faced by each type of implementation.

Table 2.1. Literature Review Matrix

Author	**Ref.#**	**Method Used**	**Gaps**
Modi et al.	[56]	This method uses Snort to capture the packets and raise alarm for known attacks, Signature Apriori algorithm is used to generate new attack signatures from partial signatures and update the snort rule. Derivatives of known attacks are also detected.	This method depends upon the knowledge about the partial signatures and cannot detect zero-day attacks.
Rizvi et al.	[59]	This method uses low interaction honeypots to generate the rules using genetic algorithm. High honeypots use fuzzy logic to detect anomalies	This model will not detect internal attack as honeypots are facing internet.
Gómez et al.	[60]	Snort integrated with anomaly preprocessor is used to model the normal traffic pattern and store it in SQL. Snort detects known attacks SQL is used to detect deviation from normal traffic pattern.	Snort needs to communicate with SQL to detect anomalies.
Rajendran et al.	[61]	This method deploys NIDS at each host and on the network. SQL is used to store the anomalies and used during subsequent detection process.	Multiple NIDS communicating with SQL will cause excessive network traffic.
Aydın et al.	[63]	This paper integrates Snort with preprocessors PHAD and NETID. PHAD and NETID model the incoming normal traffic at the protocol level like TCP, UDP, ICMP, FTP etc. and raise an alarm if any deviation is found.	Both PHAD and NETID raise alarm once only for n number of similar deviations which can be misleading.
Uddin et al.	[64]	Main complementary database updates multiple small signature databases with a subset of frequently occurring attack signatures	Delay in updating the small signature database can cause detection to fail
Kim et al.	[65]	This method uses Decision Tree to categories the data into normal, known and unknown attacks. Unknown attacks are fed to one class support vector machine which is trained on multiple normal data sets from Decision tree for further classification.	Decision Tree can become complicated for high dimensional data.
Chiba et al.	[66]	Snort and Optimized Backpropagation Neural Network	Delay in updating the database can result in Intrusions.

Jabez & Muthukumar	[67]	This method uses Snort to collect network traffic and detect known attacks. Anomaly data is fed to K-NN to detect the outliers. K-NN is trained on data from distributed environments to increase accuracy,	K-NN can be computationally expensive for high dimensional data.
Teng et al.	[68]	In this method event detector categorizes the events based on protocol like TCP, UDP and ICMP. Multiple SVMs are multiple SVMs are then used to detect the anomalies for each protocol.	It will give incorrect results in case of network traffic bursts.
Zhang et al.	[70]	This method used the concept of Artificial Immune system agents for Intrusion detection system.	Space complexity increases with increases in antigens/
Afzali Seresht and Azmi	[71]	This method used the concept of Artificial Immune system agents for Intrusion detection system.	Increase in agents requires more communication.
Teng et al.	[72]	This method uses Decision Tree to categories the data into normal, known, and unknown attacks. Unknown attacks are fed to one class support vector machine which is trained on multiple normal data sets from Decision tree for further classification.	Performance of the agents need to be monitored.
Idrissi et al.	[73]	This method used Agent for detection of the attacks and encryption was used for communication in the network environment.	In case of Cloud, single agent may result in delay in detection
Toumi et al.	[75]	It uses multiple agents to detect the anomalies at each VM. Detected anomaly is stored in SQL. This information is fetched by static agents and communicated to other agents. Rules in snort at each VM are updated. Firewall is then updated by these rules	Updating database, VM and firewall is time consuming. Agent performance also needs to be monitored.
MacDermott et al.	[78]	In this method the cloud broker from multiple federated clouds communicate with each other to make a decision about the attack. Dempster-Shafer theory is used to fuse the alarms	Communication between clouds needs to be secure. Correlation will be time consuming.
Huh	[80]	Virtual Machine Introspection is used to detect anomalies at each cloud domain. Correlation is used to correlate all the alerts from all the domains update each domain with new anomalies,	Model is complex as multiple domain communication can cause delay in correlating alerts

Fung & Zhu	[81]	This method uses multiple NIDS from different vendors to detect different types of attacks. NIDSs collaborate with each other to detect anomalies. Each NIDS maintains a list of all other NIDSs.	Consulting multiple NIDS can cause delay.
Ficco et al.	[82]	This method proposes a layered model of detection method. If the NIDS at the domain layer is not able to make decision, alarms are correlated with the security engine in the federated clous.	Communication overhead due to multiple layers of communication.
Idhammad et al.	[83]	In this method traffic at each cloud edge server is classified using naïve bayes algorithm. Suspicious traffic is sent to central database for attack classification by Random Forest tree.	Will not detect Internal attacks as the traffic id monitored om edge server.
Abusitta et al.	[84]	In this method Denoising autoencoder is used to constrict full feedback from partial of other clouds. DA is used to train the Deep Neural network for classification.	Auto encoders are Computationally expensive
Zhang et al.	[85]	This method uses Many Objective Optimization algorithm to optimize the number of features and to improve the accuracy. K-NN is used to classify the data into normal and anomaly.	Number of features is still high. KNN can be computationally expensive in case of high dimensional data.
Samriya et al.	[86]	This method uses Fuzzy cluster means with Spider Monkey Algorithm to select the most relevant features. Artificial Network was used for the classification of data,	In case of high dimensional data clustering can be time consuming
Rabbani et al.	[87]	This method uses PCA for the dimensional reduction and Particle Swam Based Probabilistic Neural Network for classification of data.	PSO has low convergence rate
Krishnaveni et al.	[88]	This method uses multiple feature selection methods to rank the featyres. Union of these features is taken as the final set of features. Majority voting bases ensemble classifier os used for detection.	Using multiple feature selection algorithms and aggregating the results will increase time complexity.
Thilagam et al.	[89]	In this method the data is converted into images, Convolutional Neural Network is then used to Classify the images. Ant Lion Optimization is used to reduce the error rate.	Convolutional Neural Network are slower and take longer time to train.

Mayuranathan et al.	[90]	Heap Optimization is used for data preprocessing. Red Deer algorithm is used to optimize the feature selection process. Finally Deep neural network is used for the classification of data.	Heap optimization is has high time complexity, so the taken to train the model is higher
Kaushik et al.	[91]	This method uses Chi-Square and information gain as the feature selection methods. Adaptive Boosting is used for classification.	AdaBoost method can overfit in noisy data
Mahmood & Agrawal	[92]	This method used Principal Component Analysis as dimensional reduction technique and Artificial Neural Network as the classifier. Backpropagation is used to adjust weights.	Neural Network with Backpropagation is computationally expensive.

Table 2.2. Challenges in different cloud NIDS Implementations

Intrusion Detection technique	**Challenges**
Hybrid Technique	(i) Snort will not be able to keep up when the traffic is high. This will result in dropping of packets by snort. (ii) With thousands of rules in snort, it will consume more computational power than simple anomaly detection.
Agent Based Intrusion Detection System	(i) Communication and Coordination between agents is a challenge. This increases as the number of agents increases. (ii) Monitoring the agents to make sure that they have performed the task is also a challenge.
Intrusion Detection for Federated Clouds	(i) Feedback from different IDS need to arrive in timely manner without any delay. (ii) Communication between IDS needs to be secure. (iii) Using encryption technique to secure the attack information can slow down detection process.
Intrusion Detection for Cloud	(i) Need training and testing data for the model. (ii) More the training data and more its variety, more will be the efficiency of the model. (iii) Selecting relevant features and avoiding the problem of "Curse of Dimensionality" is a challenge.

While performing the literature review, some papers which focused on ANIDS solutions for the cloud computing were selected for further study as shown in table 2.3. The performance of these ANIDS models form the basis of the comparison of our work. The type of the ANIDS and the performance metrics used by these papers is also shown in this table.

Table 2.3. Research Papers and their Performance Metrics that were considered for further study

Ref.	**IDS type**			**Evaluation Metrics Used**								
	HIDS	*NIDS*	*HyIDS*	*Accuracy*	*TPR*	*FPR*	*TNR*	*FNR*	*ROC*	*Precision*	*Recall*	*F1*
[83]		√		√	x	√	x	x	√	x	x	x
[84]		√		√	x	x	x	x	x	x	x	x
[85]		√		√	x	x	x	x	x	√	√	√
[86]		√		√	x	x	x	x	x	√	√	√
[87]		√		x	√	√	x	x	x	√	x	√
[88]		√		√	√	√	x	x	x	x	x	x
[89]		√		√	√	√	x	x	x	√	√	√
[90]		√		√	√	x	√	x	x	√	√	√
[91]		√		√	√	√	√	√	x	√	√	√

2.6. Research Gaps

While going through the literature review, following gaps were identified.

(a) Simultaneous Features and Accuracy optimization

ANIDS needs to process every feature of the network packet to identify whether the packet is a normal or anomaly. Processing all the features is an overhead for a cloud ANIDS. Hence this computational time needed to be reduced to increase the detection performance. During the literature review, we found that most of the focus was given to the accuracy of ANIDS but using more features [83-91]. Limited effort was done to optimize the features and accuracy at the same time. Even using more features didn't yield the desired results. There were two aspects of the problem i.e. to reduce the feature set and at the same time increase the accuracy of ANIDS. We treated this problem as a muti-objective optimization problem, where we needed to maximum the accuracy and minimize the feature set. Thus, we formulated a solution by optimizing the different components of ANIDS and then integrating them together to develop an optimized model.

(b) Removing Noise and Collinearity

While eliminating the redundant features in the dataset is important to reduce the noise and increase the accuracy, it is also important to reduce the collinearity between the independent variables and transform them into a fewer variable, The earlier literature used either the feature selection methods or feature extraction methods [85-88]. This could either remove the redundant features or transform the features but not both. We identified an opportunity to use both methods to get better results. Hence, we used optimized feature selection method followed by optimized feature extraction method to get rid of redundancy and collinearity.

(c) Optimizing Accuracy

To evaluate the performance of ANIDS, confusion matrix parameters like accuracy, TPs, TNs etc. are used. Optimizing these parameters is one of the objectives of developing an efficient ANIDS. During the literature review, we identified that efforts were made to improve these parameters using various MLTs [85-91]. However, for ANIDS, accuracy needs to be as close to 100% as possible. The work that was done so far focused on improving the accuracy using a single classifier. Some researchers used multiple ensembled classifiers but not in a layered manner. We identified the scope to use multiple classifiers in series or layers to process the network packets at multiple layers which improved the performance of ANIDS.

2.7. Conclusion

Network intrusion detection systems can be categorized based on their implementation as Hybrid, Collaborative based, Agent based and intrusion detection systems for cloud computing. Cloud based intrusion detection systems have mainly used some optimizing techniques like spider monkey algorithm and principal component analysis for dimensional reduction and classifiers like deep learning, clustering techniques and naïve bayes algorithm etc. far classification of data. However, we are far from the accuracy of

100% which is still a topic of research. Another gap is that the number of features used to get the present level of accuracy is too high which needs to be optimized. Having more features for network intrusion detection increases the time complexity as well as makes the model more complex and computationally expensive. For NIDS to be implemented in cloud computing, it must be able to make accurate decisions by inspecting fewer features of the network packets. This means that we need to reduce the dimensionality of the data and at the same time increase the accuracy of network intrusion detection system. Hence, this two-dimensional problem of accuracy vs number of features has emerged as main research question from our literature review.

Chapter 3

CHAPTER 3 : PERFORMANCE EVALUATION OF FEATURE SELECTION METHODS

3.1. Introduction

Every dataset has some features that reflect the category of each record in the dataset. Network traffic also has some features like source IP address, destination IP address, source port etc. For an ANIDS to make accurate decisions, it must inspect all these features and analyze their relationship with the class variable (dependent variable). This is done to find out how strongly each of these features is related to the class variable or in other words what is the magnitude and direction of impact that each of the independent variables has on the class variable. More strongly a feature is related to the class variable, greater the impact it has on the class variable. Strongly related features will have greater weightage in classification of network data. Thus, we needed to find those features in a dataset that have greater impact on the class variable. Other features which have less or no impact on the class variable are termed as irrelevant and are removed from the dataset. In this chapter we evaluated four filter-based feature selection algorithms i.e Correlation Based feature Selection with Subset Evaluation (CfsSubsetEval), Correlation Based Attribute Evaluation (CorrelationAttributeEval), Information Gain Based Attribute Evaluation (InfoGainAttributeEval) and Relief Attribute Evaluation (ReliefAttributeEvual). One of the advantages of filter-based feature selection method is that they are independent of any classifier and are faster than wrapper methods. However, wrapper methods take into consideration the classifier hypothesis and can handle feature dependencies. The aim of this chapter was to identify filter-based feature selection algorithm that would select those features which yield best accuracy. The selected method for feature selection was used in the final proposed model.

Feature selection algorithms have applications in the real world. They can be used to increases the performance of algorithms by accelerating them and makes the model more simpler [46]. By reducing the features, we can reduce the time complexity of the algorithm. In case of ANIDS, using less features can increase the detection rate of the intrusions and thus increase the efficiency of ANIDS. This is especially true in case of an ANIDS for cloud computing. Features selection processes fall into the broader category of feature engineering.

3.2. Feature Engineering Process

Feature engineering is the process of selecting the relevant features, transforming the existing features into a new feature set, scaling the features, or encoding the features so that they can be used with the machine learning algorithm [93]. The goal of feature engineering is to improve the accuracy of the machine learning algorithm by selecting only those features which are highly co-related with the class feature [94]. Figure 3.1 shows the life cycle phases of ANIDS model creation.

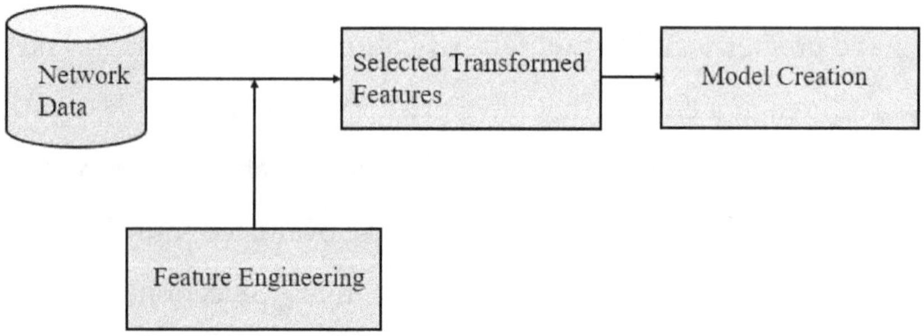

Fig.3.1. Feature Engineering in ANIDS Model Creation

Feature engineering is a broader term which encompasses the manipulation of features of the dataset. Feature engineering is performed prior to the model creation process and after the sample dataset is imported into the model creation tool. Classification of featuring engineering techniques based in their function is shown in figure 3.2. The main objective of the feature engineering is to increase the performance of the predictivity model by transforming the feature space [95].

Feature extraction and feature selection techniques fall under the category of dimensional reduction. Feature selection is used to eliminate the irrelevant features from the data to increase accuracy and feature extraction, on the other hand removes the collinearity between the independent variables by combining those variables which are highly collinear. Hence the original dataset is transformed into a new dataset with less features.

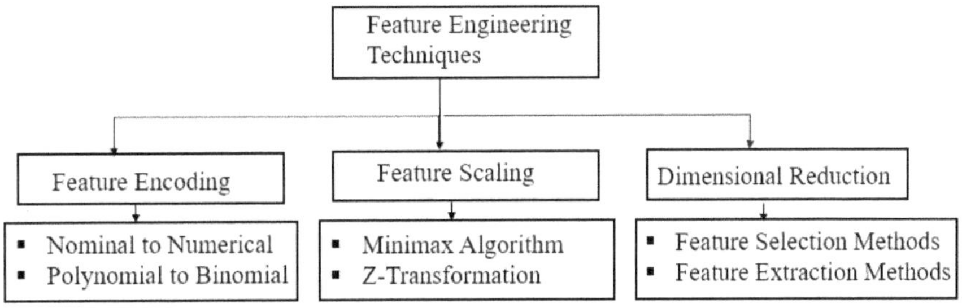

Fig. 3.2. Feature Engineering Techniques

3.3. Methodology Used for Feature Selection Evaluation

Figure 3.3 shows the methodology that was used in evaluating four feature selection methods. We used NSL-KDD dataset to train and test the model and SVM as the classifier to build the model. We built four different models using the features selected by four feature selection methods. SVM based classifier was then used to classify the data using the selected features. Accuracy, false alarms, and other confusion matrix parameters were observed and analyzed to arrive at a conclusion.

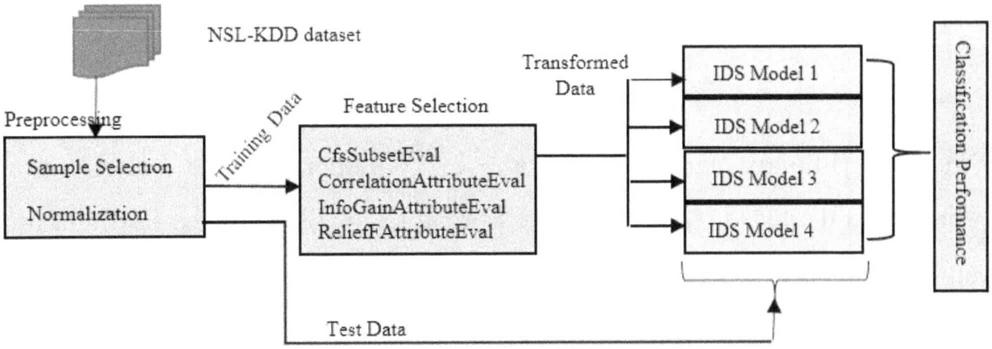

Fig. 3.3. Feature Selection Model

3.3.1. Sample Dataset Selection

We used NSL-KDD dataset for performing our experiment. This dataset contains KDDTrain+.ARFF file which has 129973 instances with 42 features including the class feature. Each of the instances is classified as "normal" or "anomaly. Out of 129973 instances, 67434 instances are labelled as “normal” and 58630 are “anomaly”. There are no redundant or duplicate instances in this file [58]. We used 70% of the data for training the model and 30% of the data for testing. Confusion matrix parameters were used to measure the performance of the model.

3.3.2. Normalization using Minimax Algorithm

Figure 3.4 shows some of the feature in the dataset with numerically high values which needed to be normalized.

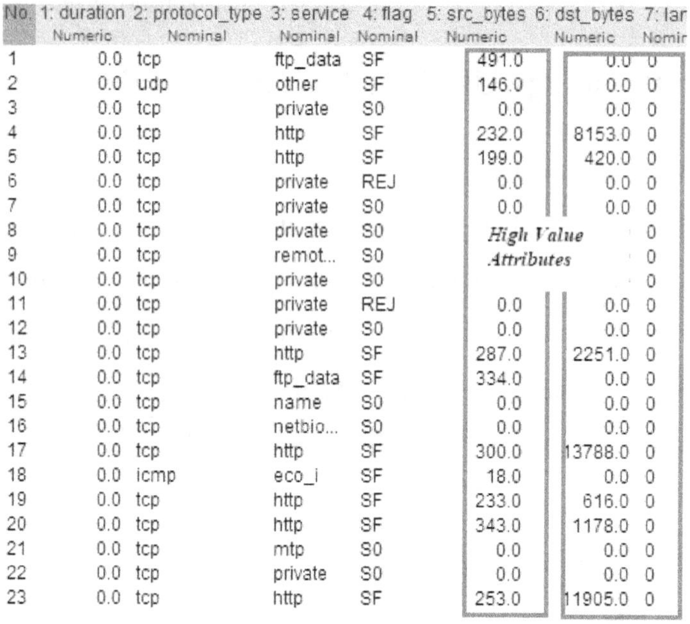

No.	1: duration Numeric	2: protocol_type Nominal	3: service Nominal	4: flag Nominal	5: src_bytes Numeric	6: dst_bytes Numeric	7: lar Nomir
1	0.0	tcp	ftp_data	SF	491.0	0.0	0
2	0.0	udp	other	SF	146.0	0.0	0
3	0.0	tcp	private	S0	0.0	0.0	0
4	0.0	tcp	http	SF	232.0	8153.0	0
5	0.0	tcp	http	SF	199.0	420.0	0
6	0.0	tcp	private	REJ	0.0	0.0	0
7	0.0	tcp	private	S0	0.0	0.0	0
8	0.0	tcp	private	S0			0
9	0.0	tcp	remot...	S0			0
10	0.0	tcp	private	S0			0
11	0.0	tcp	private	REJ	0.0	0.0	0
12	0.0	tcp	private	S0	0.0	0.0	0
13	0.0	tcp	http	SF	287.0	2251.0	0
14	0.0	tcp	ftp_data	SF	334.0	0.0	0
15	0.0	tcp	name	S0	0.0	0.0	0
16	0.0	tcp	netbio...	S0	0.0	0.0	0
17	0.0	tcp	http	SF	300.0	13788.0	0
18	0.0	icmp	eco_i	SF	18.0	0.0	0
19	0.0	tcp	http	SF	233.0	616.0	0
20	0.0	tcp	http	SF	343.0	1178.0	0
21	0.0	tcp	mtp	S0	0.0	0.0	0
22	0.0	tcp	private	S0	0.0	0.0	0
23	0.0	tcp	http	SF	253.0	11905.0	0

Fig. 3.4. Features with High Numeric Values in the Dataset

We used Minimax algorithm to scale every numerical value between 0 and 1. This eliminated the bias in the results due to features of high values. Formula for mini-max algorithm is given by equation 3.1 [96].

$$X_{new} = \frac{X - \min(X)}{\max(X) - \min(X)} \qquad \text{(3.1)}$$

where X is the value of the feature to be normalized, min(X) is the minimum numerical value of this feature, max(X) is the maximum numerical value of the feature and X_{new} is the new normalized value of the feature which replaces the value of X.

3.3.3. Feature Selection Methods used for Evaluation

We selected four features selection methods for evaluating their performance on SVM classifier.

(a) Correlation Based Feature Selection with Subset Evaluation (CfsSubsetEval)

This correlation-based feature selection method calculates the correlation between each individual feature and the class feature. At the same time, it also checks the intercorrelation between individual features [23]. The correlation between a feature and the class feature gives the predictive ability of the feature. The subset of features which are highly correlated with the class feature and at the same time have low interrelationship are selected.

(b) Correlation Based Attribute Evaluation (CorrelationAttributeEval)

Like CfsSubsetEval, this method also evaluates the importance of a feature based on its correlation with the class feature. However, it does not take intercorrelation between features into consideration.

(c) Information Gain Attribute Evaluation (InfoGainAttributeEval)

Information gain is used to explain the characteristics of a dataset and identify the features that are most related to the class feature. Information gain uses entropy to decide the importance of a feature. Entropy measures the uncertainty present in the data items [97]. Entropy of a dataset X is given by equation 3.2 [98].

$$H(X)=\sum P(x_i)I(x_i)=-\sum P(x_i)\log_b P(x_i) \quad \text{.....................(3.2)}$$

where $P(x_i)$ is the probability that x_i is in X. Information gain for any attribute A is given by equation 3.3.

$$\text{Gain}(A)=\text{Info}(D)-\text{Info}_A(D) \quad \text{...........................(3.3)}$$

where,

$$\text{Info}_A(D) = \sum_{j=1}^{v} \frac{|D_j|}{|D|} \times \text{Info}(D_j)$$

Where A is a feature in the dataset D.

(d) Relief Attribute Evaluation (ReliefFAttributeEval)

This feature selection method is instance-based. It samples instances randomly and checks neighboring instances of the same and different classes. It operates on discrete and continuous class data. Parameters specify the number of instances to sample, the number of neighbors to check, whether to weight neighbors by distance, and an exponential function that governs how rapidly weights decay with distance.

3.4. Experimental Results

We used NSL-KDD dataset to evaluate the accuracy of feature selection methods on SVM. This dataset has some high value features which were normalized using minimax algorithm. The normalized data now had 41 features and one class variable.

Each of these feature selection algorithms ranked the features based on the predicting power. We selected top 10 features from each of the feature selection algorithms as shown in the table. 3.1. Each of these feature sets were then used to evaluate the accuracy of the model using SVM classifier. The results of the confusion matrix using each of the feature selection algorithms is shown in the table 3.2.

Table 3.1. Features Selected by Feature Selection Algorithms

S. No	CfsSubsetEval	CorrelationAttributeEval	InfoGainAttributeEval	ReliefFAttributeEval
1	dst_bytes	same_srv_rate	flag	service
2	flag	dst_host_srv_count	same_srv_rate	flag
3	logged_in	dst_host_same_srv_rate	dst_host_srv_count	same_srv_rate
4	diff_srv_rate	logged_in	dst_host_same_srv_rate	dst_host_count
5	src_bytes	dst_host_srv_serror_rate	logged_in	dst_host_srv_serror_rate
6	srv_serror_rate	dst_host_serror_rate	dst_host_serror_rate	dst_host_srv_count
7	dst_host_srv_diff_host_rate	flag	dst_host_srv_serror_rate	dst_host_serror_rate
8	same_srv_rate	serror_rate	serror_rate	logged_in
9	dst_host_srv_serror_rate	srv_serror_rate	count	dst_host_same_src_port_rate
10	service	count	srv_serror_rate	srv_serror_rate

It can be seen from the table 3.2 and the figures 3.5 to figure 3.7 that CfsSubsetEval and ReliefAttributeEval have performed almost equally well on all performance parameters. As compared to other features selection methods these two methods yielded improved performance.

Table 3.2. Performance of Feature Selecti Algorithms

Method used /Accuracy	Performance	TPR	FPR	Precision	Recall	F-Measure	ROC
	Best	0.965	.040	0.965	0.965	0.965	0.963
CfsSubsetEval	*Worst*	0.960	0.035	0.960	0.960	0.963	0.940
0.962	*Avg.*	**0.963**	**0.038**	**0.963**	**0.963**	**0.963**	**0.945**
	Best	0.955	0.118	0.902	0.955	0.928	0.918
CorrelationAttributeEval	*Worst*	0.882	0.045	0.945	0.882	0.912	0.918
0.920	***Avg.***	**0.921**	**0.084**	**0.922**	**0.921**	**0.921**	**0.918**
	Best	0.960	0.046	0.960	.960	.960	0.957
InfoGainAttributeEval	*Worst*	0.954	0.040	0.954	0.954	0.954	0.957
0.957	***Avg.***	**0.957**	**0.043**	**0.957**	**0.957**	**0.957**	**0.957**
	Best	0.966	0.040	0.965	0.966	0.966	0.963
ReliefFAttributeEval	*Worst*	0.960	0.034	0.961	0.966	0.961	0.963
0.963	***Avg.***	**0.964**	**0.037**	**0.964**	**0.964**	**0.964**	**0.963**

From figure 3.5, it is evident that the average accuracy of CfsSubsetEval and ReliefAttributeEval is almost the same with some negligible difference. However, from the table 3.2, ReliefAttributeEval has a little bit of better accuracy by 0.001 (0.1%). This means out of 1000 Intrusion ReliefAttributeEval will be able to detect one intrusion more than the CfsSubsetEval.

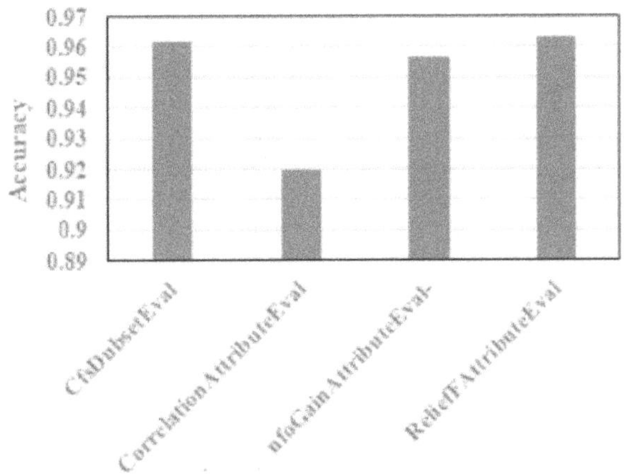

Fig.3.5. Performance on Accuracy

From figure 3.6 we again see that the TPR of both CfsSubsetEval and ReliefAttributeEval are same as in case of accuracy with a minor difference of 0.001.

This means that out of 1000 true intrusion that occur, ReliefAttributeEval will be able to detect one more than CfsSubsetEval.

Figure 3.7 shows the performance on FPR using the four feature selection methods. It is clear from this figure that ReliefAttributeEval has performed little bit better than CfsSubsetEval. However, the difference in the performance is again negligible. It can be safely said that these two perform equally well on FPR parameter.

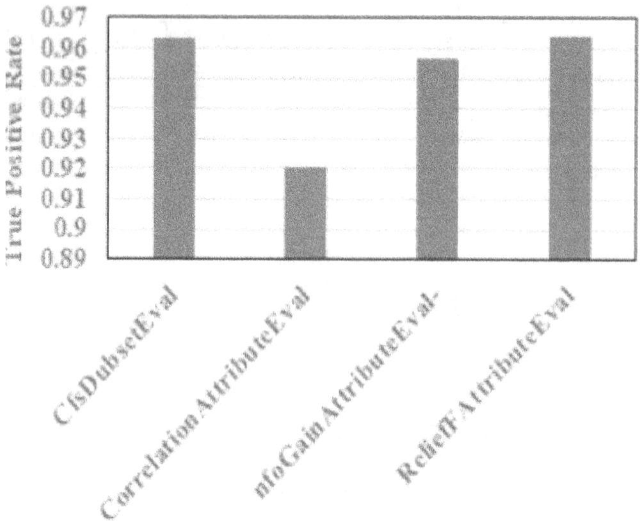

Fig.3.6. Performance on True Positive Alarms

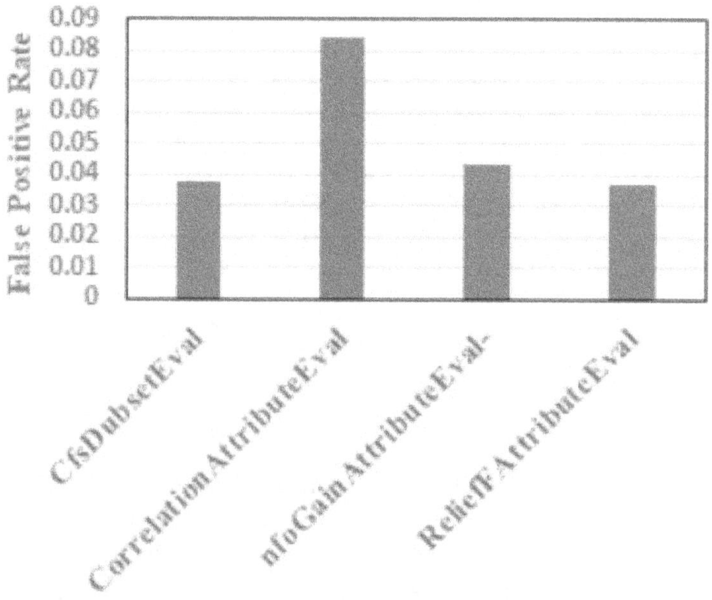

Fig.3.7. Performance on False Positive Alarms

3.5. Conclusion

Feature selection is one of the important tasks of building an intrusion detection system model. Various Feature selection methods have been used by researchers to increase the accuracy of an intrusion detection system. In fact, the efficiency of an intrusion detection system depends upon how efficiently a feature selection method can extract the most relevant features from the dataset and rank them. In this chapter we have compared four feature selection methods and used support vector machine to test the classification results using NSL-KDD dataset. The results showed that CdsSubsetEcal and ReliefAttributeEval performed equally well on all performance metrics as compared to other feature selection methods.

Chapter 4

CHAPTER 4: A COMPARATIVE STUDY OF FEATURE EXTRACTION TECHNIQUES FOR NETWORK INTRUSION DETECTION SYSTEM IN CLOUD COMPUTING

4.1. Introduction

NIDS is an effective tool for detecting the Intrusions in a network. It must analyze all the features of a network packet to find any abnormality in the packet and raise an alarm if any suspicious packet is found. Since cloud computing is characterized by a high volume of incoming traffic, which is high dimensional in nature, examining all the network packets can exhaust the NIDS. This, in turn can cause NIDS to drop some of the packets. This can be a security risk for cloud resources including critical infrastructure as well as the critical data stored in the cloud environment. Any anomaly in the network traffic which is not detected in timely manner can be a cause of cyber-attack. To prevent this, NIDS should be able to make accurate decisions by examining the fewer features of a packet. Feature extraction techniques can be used to transform the original features set into a new set of features which are less in number as compared to original feature set, thus reducing the dimensionality of the dataset. Some of the techniques that are used for this purpose include ICA, PCA and SVD. Each of these feature extraction techniques use statistical means to reduce the feature set in a data. For our work, we needed to identify the feature extraction technique that would give better accuracy for classification. For this purpose, we selected three above mentioned feature extraction techniques and compared their performance. Hence, in this Chapter, we developed three different classification models using the features that were extracted by these three feature extraction techniques. The classification was done on three different classifiers to arrive at a conclusion. We used NSL-KDD dataset for this work . We observed that features extracted by SVD technique produced more consistent accuracy in classification on different classifiers as compared to other two feature extraction techniques.

4.2. Methodology Used

Figure 4.1 shows the methodology that was used to compare the performance of ICA, PCA and SVD using ANN, RF and NBA as the classifiers. As can be seen from the figure, we have encoded the NSL-KDD dataset followed by the normalization process to scale the variables. To get better results, we also removed the irrelevant features which could have impacted the accuracy and given misleading results. For comparative analysis, we used three classifiers, and each classifier was evaluated on features extracted by all the three feature extraction techniques individually. To get reliable results, we repeated the experiment with different seed values and took multiple readings. Performance analysis was done based on the average accuracy, average false alarms, average number of features and also the consistency with which these results were obtained.

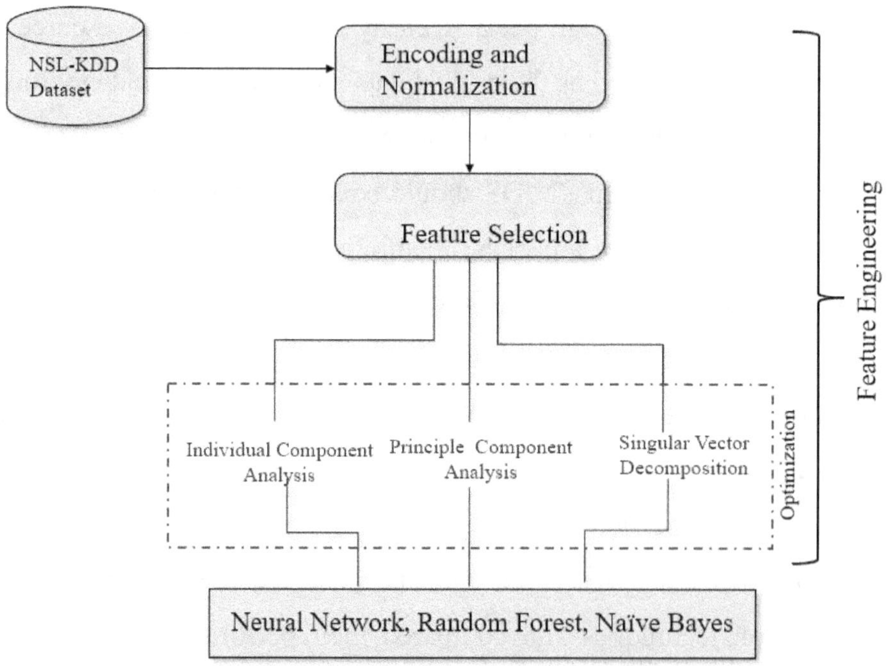

Fig.4.1. Methodology used for Comparison

4.2.1. Dataset Selection

The main reason for using the dataset was to evaluate the performance of NIDS model developed by using three feature extraction methods. Various datasets like *NSL-KDD, CAIDA, ISCX , CICIDS* etc. are publicly available. These datasets contain the malicious as well as normal data packets for training and testing NIDS models [33]. Out of these datasets, NSL-KDD dataset and KDD cup datasets are most widely used in network domain by researchers [43]. The fact that NSL-KDD dataset does not contain any redundant and duplicate entries, it is safe to say that the NIDS trained on this dataset will not give biased results. Hence, we used NSL-KDD dataset for our work. The description of this dataset is shown in the table 4.1. Each example in the dataset has 41 features apart from the class label. Out of 125973 examples present in the dataset, we randomly selected 9448 examples for our work. Our sample included 4410 examples which were labelled as anomalies and 5028 examples were labelled as normal. Hence sample dataset was well balanced. In the sample dataset, 70% of the examples were used for training purpose and 30% were used for testing purpose.

Table 4.1. Description of NSL-KDD Dataset

Feature	Type	Feature	Type	Feature	Type
duration	Integer	su_attempted	Integer	same_srv_rate	Real
protocol_type	Polynomial	num_root	Integer	diff_srv_rate	Real
service	Polynomial	num_file_creations	Integer	srv_diff_host_rate	Real
flag	Polynomial	num_shells	Integer	dst_host_count	Integer
src_bytes	Integer	num_access_files	Integer	dst_host_srv_count	Integer
dst_bytes	Integer	num_outbound_cmds	Integer	dst_host_same_srv_rate	Real
land	Integer	is_host_login	Integer	dst_host_diff_srv_rate	Real
wrong_fragment	Integer	is_guest_login	Integer	dst_host_same_src_port_rate	Real
urgent	Integer	count	Integer	dst_host_srv_diff_host_rate	Real
hot	Integer	srv_count	Integer	dst_host_serror_rate	Real
num_failed_logins	Integer	serror_rate	Real	dst_host_srv_serror_rate	Real
logged_in	Integer	srv_serror_rate	Real	dst_host_rerror_rate	Real
num_compromised	Integer	rerror_rate	Real	dst_host_srv_rerror_rate	Real
root_shell	Integer	srv_rerror_rate	Real	class	Polynomial

4.2.2. Feature Engineering Techniques

We used various feature engineering techniques to manipulate the dataset and make it compatible with the machine learning algorithms. This includes feature encoding, normalization, and feature selection prior to using feature extraction methods.

a) Feature Encoding

Since the sample dataset consisted of some nominal and polynomial features, we needed to transform these features into numeric type to make dataset compatible with MLT. Table 4.2 shows the description of nominal and numeric features. Hence, Prior to normalization, nominal and polynomial features were encoded to the numeric values.

Binomial features in the dataset were encoded to 0 and 1. Thus we had a transformed dataset with 41 numeric features and one class variable.

Table 4.2. Type of Features in NSL-Dataset

Feature Type	Number of features	Encoding needed
Numeric	34	X
Nominal	8	√

b) Z-Transformation

Z-transformation also called the statistical transformation was used to normalize the data. Normalization is an important task in feature engineering to avoid the bias towards the features having higher values. This method does not alter original distribution of the data and is not influenced by outliers. In this method mean of the data is calculated and is subtracted from each value which is then divided by standard deviation as shown in equation 4.1. [37]

$$Z_i = (mean(x) - x) / S \quad \text{.................} \quad (4.1)$$

Where Z_i is the transformed value, *mean(x)* is the mean value of the numeric range, x_i is the numeric value to be normalized and S is the standard deviation. Due these features, Z-transformation method was used for normalizing the numeric data [99].

c) Feature Selection using Correlations Based Feature Selection

Feature selection was done using Weka machine learning tool. We used correlation-based feature selection as it measures the inter-correlation between the independent variables in addition to the correlation with the class variable while selecting the features. This method falls into the category of filter-based methods and uses statistical techniques to select only those features which are highly correlated with the dependent or class variable and have very low collinearity. This method is independent of any classifier unlike wrapper methods [74]. Since we wanted the feature selection method not to be influenced by classifiers, we chose correlation-based feature selection method for our work.

This method uses Person's correlation coefficient to select the relevant features for classification [5],[6]. Person correlation coefficient is given by equation 4.2.

$$r_{xy} = \frac{\sum_{i=1}^{n}(x_i - \bar{x})(y_i - \bar{y})}{\sqrt{\sum_{i=1}^{n}(x_i - \bar{x})^2}\sqrt{\sum_{i=1}^{n}(y_i - \bar{y})^2}} \quad \text{.......... (4.2)}$$

During the feature selection process we used PSO method to optimize the search of feature space [87]. In case of features selection, features were considered as the particles in the feature space. PSO works by evaluating the objective function (Accuracy) at its current location and decides the next movement [100]. The next movement is decided based on fitness value of next location and the current location. This process continues and the flock or swarm (features) moves close to the solution or the optimum value of fitness function.

We used correlation-based feature selection method with 20-fold cross validation to evaluate the accuracy of features on deep learning model. PSO algorithm is shown in Algorithm 4.1, where $\bar{x}$ is the current position of the particle in D dimensional search space, pi is the previous best position, mv_i is the velocity, pop_t stores the optimized result after each iteration. Table 4.3 shows the features that were selected by feature selection algorithm. There were some features which showed number of folds as 0%. This means that these features were not considered relevant and were never used in the evaluation process by correlation algorithm. Hence, we excluded such features from our dataset and considered only those features which showed number of folders as greater than zero. Hence in the new transformed dataset, we now had 26 features including the class feature.

Table 4.3. Features Selected using Feature Selection Method

duration	num_file_creations	num_compromised	dst_host_diff_srv_rate
protocol_type	num_shells	su_attempted	dst_host_same_src_port_rate
service	num_access_files	num_root	dst_host_rerror_rate
dst_bytes	num_outbound_cmds	srv_count	dst_host_count
land	is_host_login	srv_rerror_rate	clas
wrong_fragment	is_guest_login	same_srv_rate	
logged_in	count	diff_srv_rate	

Algorithm 4.1. Particle Swarm Optimization Algorithm

1. Initialize the population in search space
2. *for* each particle find the desired optimum fitness function.
3. ***start loop***
4. compare the fitness of particle with $p_{opt.}$
5. If the current value is better than p_{opt}
6. ***start loop***
7. set p_{opt} equal to current value
8. set p_i to the current location $\bar{x}$
9. ***end loop.***
10. identify the most optimized particle in neighborhood and assign its index to variable g.
11. modify the velocity and position of the particle using equation.
$$\begin{cases} v_i \leftarrow v_i + U(0, \phi_1) \otimes (p_i - x_i) + U(0, \phi_2) \otimes (p_g - x_i), \\ x_i \leftarrow x_i + v_i. \end{cases}$$
12. if criteria is met exit the loop
13. **end loop.**

d) Feature Extraction Techniques

Feature Extraction Techniques were used to further reduce the dimensionality of the data by removing the collinearity between the independent features and combining those features which have high collinearity [27].

Independent Component Analysis

ICA is a very general-purpose statistical technique in which observed random data are linearly transformed into components that are maximally independent from each other, and simultaneously have "interesting" distributions. Such a representation are able to capture the essential structure of the data in many applications, including feature extraction. ICA is used for revealing hidden factors that underlie sets of random variables or measurements. ICA is superficially related to PCA and factor analysis. ICA is a much more powerful technique, however, capable of finding the underlying factors or sources when these classic methods fail completely. It is parallel, distributed, computationally simple, and requires little memory space [77].

ICA is a statistical technique that is used to separate multivariate data into independent components. The goal of ICA is to find a linear transformation of data in such a way that the new transformed data is as statistically independent as possible. If observed data N is transformed into independent C components using linear transformation given by matrix M, then the equation 4.3 should satisfy.

$$C= MN \qquad \text{.........................(4.3)}$$

where M is the transformed matrix. The goal of ICA is to transform the hidden components of M in such that the resulting components are as independent as possible.

Singular Vector Decomposition

SVD is based on a principle that a matrix can be broken down into three different matrices. These include an Orthogonal matrix U, a diagonal matrix S and transpose of orthogonal matrix V. If H is a matrix with dimensions (m, n) then equation for SVD is given by equation 4.4 [101].

$$H=USVT \qquad \text{......... (4.4)}$$

Where $U^TU = V^TV=I_n$

Subject to condition $S= diag(a_1.........a_n)$

Where $a_i>0$ *for* $1\leq i \leq r$ *and* $a_j=0$ *for* $j \geq r+1$

SVD can be used for the dimensional reduction and performs well as compared to self-organizing maps [101]. Due to the performance of SVD, we selected it as one of the dimensional reduction techniques for performance comparison.

Principle Component Analysis

PCA technique was introduced by the mathematician Karl Pearson in 1901 [57]. It works on the condition that while the data in a higher dimensional space is mapped to data in a lower dimension space, the variance of the data in the lower dimensional space should be maximum. PCA is a statistical procedure that uses an orthogonal transformation that converts a set of correlated variables to a set of uncorrelated variables [34]. PCA is the most widely used tool in exploratory data analysis and in machine learning for predictive models to reduce the dimensionality of data.

The main goal of PCA is to reduce the dimensionality of a dataset while preserving the most important patterns or relationships between the variables without any prior knowledge of the target variables. PCA is used to reduce the dimensionality of a data set by finding a new set of variables called Principal Components which are smaller than the original set of variables [102]. The principal components are linear combinations of the original variables in the dataset and are ordered in decreasing order of importance. The total variance captured by all the principal components is equal to the total variance in the original dataset [45].The first principal component captures the most variation in the data, but the second principal component captures the maximum variance that is orthogonal to the first principal component, and so on. Steps that are involved in PCA include:

(i) Standardization

In this step, we need to standardize the dataset to ensure that each variable has a mean of 0 and a standard deviation of 1. Formula for standardization is given by equation 4.5.

$$Z = \frac{X - \mu}{\sigma} \qquad \text{.........................(4.5)}$$

Where μ is the mean of independent features. Standard deviation of independent features is given by equation 4.6.

$$\mu = \{\mu_1, \mu_2, \cdots, \mu_m\} \qquad \text{..............(4.6)}$$

were,

$$\sigma = \{\sigma_1, \sigma_2, \cdots, \sigma_m\}$$

(ii) Computing Covariance Matrix

Covariance measures the strength of joint variability between two or more than two variables. In other words, it indicates how much they change in relation to each other. Covariance is calculated by using the equation 4.7.

$$cov(x1, x2) = \frac{\sum_{i=1}^{n}(x1_i - x1)(x2_i - x2)}{n-1} \qquad \text{.................(4.7)}$$

value of Covariance can be positive, negative or zero. Covariance is taken as positive if the direction of change of both the variables is same. In other words, if increase in x_1 results in increase in x_2, then covariance is taken as positive. If increase in x_1 results in decrease in x_2, then covariance is taken as negative. If there is no direct relation between x_1 and x_2, then covariance is zero.

(iii) Computing Eigenvalues and Eigenvectors to identify Principal Components

Eigenvalue and Eigenvector are the scalar and vector quantities which are associated with a matrix. They are used for the linear transformation of matrix. Eigenvector and Eigenvalue does not change during the transformation of a matrix [38].

Let A be a square nxn matrix and X be a non-zero vector whish satisfies the equation 4.8.

$$AX = \lambda X \quad \text{..............................(4.8)}$$

for some scalar values λ, then λ is known as the eigenvalue of matrix A and X is known as the eigenvector of matrix A for the corresponding eigenvalue. We can also write it as in equations 4.9 and 4.10.

$$AX - \lambda X = 0 \quad \text{..............................(4.9)}$$

$$(A - \lambda I)X = 0 \quad \text{..............................(4.10)}$$

where I is the identity matrix of the same shape as matrix A. The above conditions will be true only if (A - λI) is non-invertible (i.e. singular matrix). This means the equation 4.11 should satisfy.

$$|A - \lambda I| = 0 \quad \text{..............................(4.11)}$$

From the above equation, we can find the eigenvalues. Hence corresponding eigenvector can be found using the equation 4.7.

4.2.3. Optimizing Feature Extraction using Genetic Algorithm

We used genetic algorithm to create the optimized features using ICA, PCA and SVD. The optimized feature set were then used for classification on ANN, RF and NBA algorithms. Genetic algorithm belongs to evolutionary algorithm class. Evolutionary algorithm is a term that is used to describe the algorithms that uses Darwin's evolutionary theory to solve complex problems. One such algorithm is genetic algorithm. Genetic algorithm is an adaptive, heuristics algorithm based in genetics and natural selection process that is used to generate high quality solutions for optimization problem [103]. Genetic algorithms search mimic the process of natural evolution to optimize the results. In our experiment, we used the genetic algorithm that uses the different evolutionary phases like Initialize, Mutation, Crossover, Evaluate and Selection [104]. Various steps that genetic algorithm used to select the optimized hyperparameters include.

a) Initialize Population and Crossover

We needed to select the initial population for crossover. During Crossover, the selected population started to select the features and evaluate the fitness value (Accuracy). Poor performing features were dropped and replaced with new features. This process continued till we got the best accuracy. We selected the initial population as 5. We used the crossover probability of 0.5% and we stopped the algorithm when there was no improvement in 5 consecutive iterations [89].

b) Mutation

Mutation was performed by flipping the attributes i.e, by setting the used attributes to unused. We configured the mutation type as Gaussian and selection type as tournament. Mutation allows to slightly change or tweak the parent information while slightly adding the new child information. This is used to get new and better solution [71].

c) Evaluate

In this step, all the features generated during crossover and mutation were evaluated using DNN. The accuracy for each set of attributes was calculated. Features that produced higher accuracy were more likely to be selected for next generation.

d) Selection

In this process, a selection mechanism was used to select the best performing features for the next generation. In our work, we have used tournament selection method. Tournament method selects the features with the best accuracy for the next generation. To optimize ICA, PCA and SVD and perform the classification with optimized features, we used 70% of the data as training asset and rest 30% of the data as the test dataset.

4.3. Experimental Setup

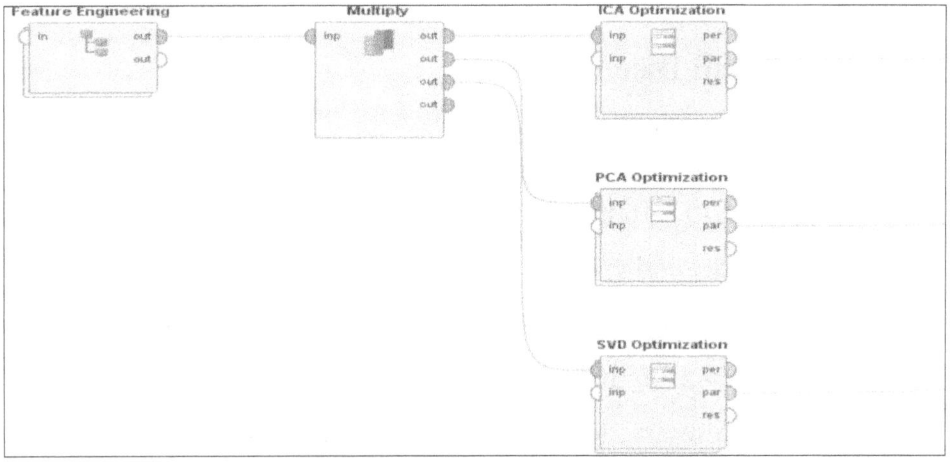

Fig. 4.2. Process Layout of Comparison Model

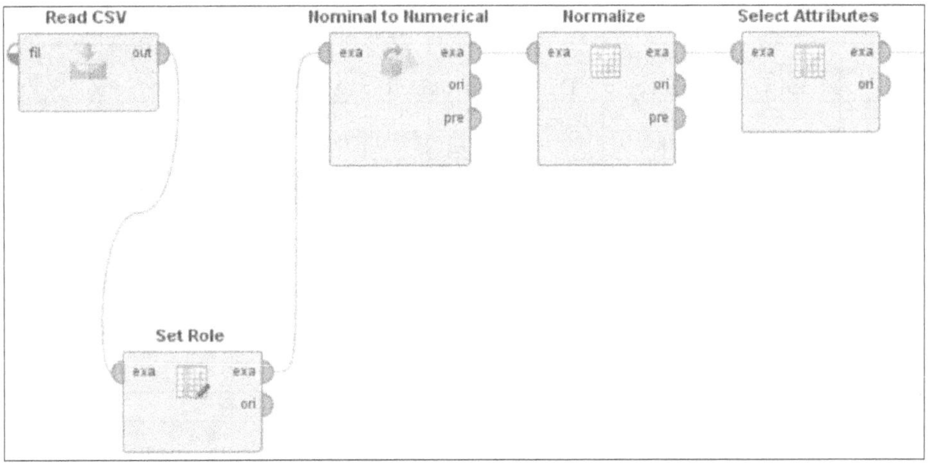

Fig. 4.3. Feature Engineering Process Layout

Experiment was conducted on a windows 10 (64 bit) machine with 16GB of RAM. Weka was used for feature selection method and RapidMiner was used for feature extraction, optimization, classification, and performance evaluation. ANN, RF and NBA algorithm modules of RapidMiner with default setting were used to gauge the accuracy and consistency in the results. Figure 4.2 shows the processes that were used for the optimization of the dimensional reduction and classification. Genetic algorithm was configured to stop after 20 iterations and the population size was set at 5. Figure 4.3 shows the feature engineering tasks that were performed using the RapidMiner.

4.4. Result and Discussion

Table 4.4 shows the accuracy performance of ANN, RF and NBA algorithms using ICA, PCA and SVD. To increase the reliability of the result, we took 5 observations for each of the model and calculated the average accuracy performance of the classifiers which is shown in the table 4.4. It can be seen from the table that the performance of average accuracy using ICA, PCA and SVD as feature extraction method varies. Even the same feature extraction technique gives different results on different classifiers. Thus, we tried to identify the feature extraction method that gives most consistent results and produced less variance in accuracy as well as the number of features extracted.

Table 4.4. Performance of Feature Extraction Techniques on Accuracy using Three Classifiers

	Neural Network			*Random Forest*			*Naïve Bayes Algorithm*		
	Seed #	Accuracy	Dimensions	Seed	Accuracy	Dimensions	Seed	Accuracy	Dimensions
	1	95.28	6	1	96.27	7	1	86.00	5
	2	95.00	5	2	95.83	5	2	86.20	2
	3	95.27	7	3	96.10	5	3	87.25	7
ICA	4	95.18	6	4	96.31	6	4	88.24	5
	5	95.40	6	5	96.25	5	5	88.89	5
	Average	**95.22**	**6**		**96.15**	**5.6**		**87.31**	**4.8**
	1	95.46	6	1	95.66	6	1	83.43	2
	2	95.18	6	2	92.77	2	2	85.70	2
PCA	3	95.28	6	3	95.06	5	3	84.76	2
	4	95.00	5	4	95.22	5	4	85.42	2
	5	95.45	6	5	95.12	5	5	84.37	2
	Average	**95.27**	**5.8**		**94.77**	**4.6**		**84.74**	**2**
	1	95.46	6	1	95.60	6	1	85.49	1
SVD	2	95.27	6	2	95.49	6	2	86.01	1
	3	95.52	6	3	95.27	6	3	86.01	1
	4	95.49	6	4	95.60	6	4	85.98	1
	5	95.55	6	5	95.16	5	5	85.55	1
	Average	**95.46**	**6**		**95.42**	**5.8**		**85.80**	**1**

4.4.1. Accuracy Performance

It can be seen from the table 4.4 that ICA produced a highest average accuracy of 96.15% using 6 features (5.6) on RF classifier and the lowest average accuracy of 87.31% using 1 feature on the NBA . Same way PCA produced a highest average accuracy of 95.27% using 6 features (5.8) and the lowest average accuracy of 84.74% using 2 features on NBA. On the other hand, SVD produced a highest average accuracy of 95.46% using 6 features and the lowest average accuracy of 85.80% using 1 feature on NBA. Since the result vary, it was necessary to check the variance in the accuracy and the dimensions to identify which feature extraction technique performed with least variance, where most of the observation were close to the average value. This determined the consistency in performance of that feature extraction technique.

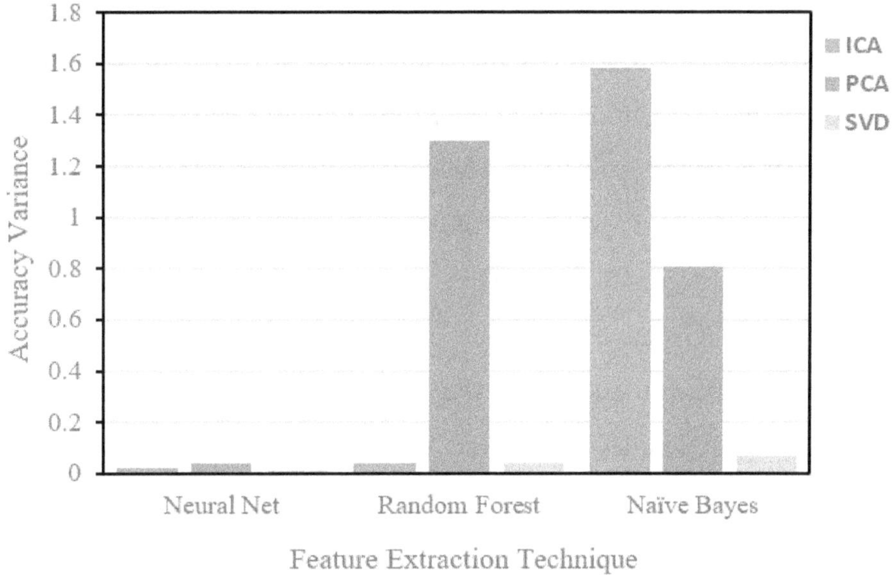

Fig.4.4. Accuracy Variance Chart

Table 4.5. Accuracy and Dimensional Variance

	Feature Extraction	Neural Net	Random Forest	Naïve Bayes
Accuracy Variance	ICA	0.02208	0.03872	1.57823
	PCA	0.03738	1.30028	0.80933
	SVD	0.01217	0.03993	0.06972
Dimensional Variance	ICA	0.5	0.8	3.2
	PCA	0.2	2.3	0
	SVD	0	0.2	0

As can be seen from the table 4.4 and figure 4.5, SVD has produced an average accuracy of 95.46% on NN, 95.42% on RF and 85.80% on NBA with a least variation as compared to ICA and PCA. Moreover, from the table 4.5, it is clear that the variance in case of average accuracy using SVD is smaller and closer to 0 as compared to other feature extraction techniques. This means that the average accuracy of SVD represents the actual and true performance of SVD. Accuracy produced by ICA and PCA have a much larger variance (closer or more than 1) and cannot produce same accuracy consistently. Thus, in terms of consistency, SVD produced better accuracy as compared to PCA and ICA.

4.4.2. Performance on Number of Features Extracted

Table 4.4 shows the average number of features that were used to get the different accuracy levels by ICA, PCA and SVD. Table 4.5 and figure 4.5 shows the variance of these features by feature extraction techniques. We can clearly see that SVD performed with least variance in features. Hence it can be concluded that the accuracy produced by classifiers by using SVD as the feature extraction technique is the consistent and represents the accuracy that has the most probability of occurring with extracted feature set. ICA and PCA have the higher dimensional variance (closer or more than 1), which means that the consistency in the number of features does not exist. Such feature engineering techniques may extract features which vary largely from the average value and can't be predicted.

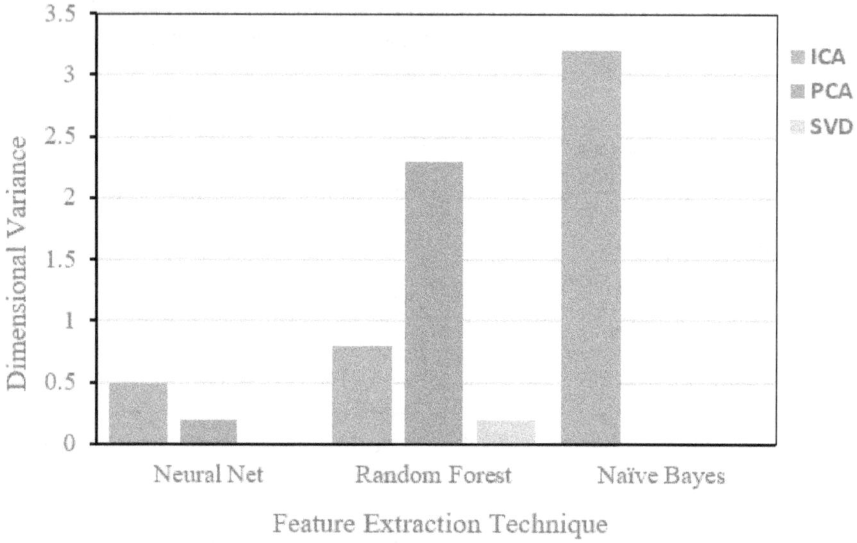

Fig. 4.5. Dimensional Variance Chart

4.5. Conclusion

Feature extraction techniques are used to reduce the dimensionality of a dataset, thus making the task of analysis much easier. Feature extraction techniques can be used in

Network Introduction Detection Systems to reduce the feature space of network traffic, thus enabling the Network Intrusion detection system to make decisions more quickly. In this paper we have done the comparative analysis of three feature extraction techniques i.e Individual Component Analysis, Principal Component Analysis and Singular Vector Decomposition using Neural Network, Random Forest and Naïve Bayes Algorithm as the classifiers. Test was performed on NSL-KDD dataset. It was observed that Singular Vector Decomposition method yielded better results with more consistency than other two feature extraction methods. Hence, we concluded that SVD is a preferred method of feature extraction technique for Network Intrusion Detection System for Cloud computing.

Chapter 5

CHAPTER 5 : RECENT APPROACHES USED IN ANOMALY BASED NETWORK INTRUSION DETECTION SYSTEMS

5.1. Introduction

Intrusion detection is basically a classification problem because the network traffic needs to be classified as "norma" or "anomaly". So, classifier is one of the main tools that is used to develop a NIDS. Dataset is important in developing the model as it helps us to train and evaluate the efficiency of the model. However, dataset may contain some redundant or irrelevant examples which need to be removed. Hence feature engineering is done prior to classification of the data. The collinearity between the independent variables is another challenge which is removed by combining such features together. This is done by using feature extraction process. While evaluating the features for their relevance, there needs to be an algorithm which can search the feature space more efficiently. Search algorithms are used in combination with feature selection algorithms to search the feature space and identify the most relevant features from a feature. Figure 5.1 shows the algorithms that have recently been used in developing the NIDS model for cloud computing [41] .

In this chapter we have briefly explained various machine learning techniques that have been used recently in developing the ANIDS models for cloud computing. These

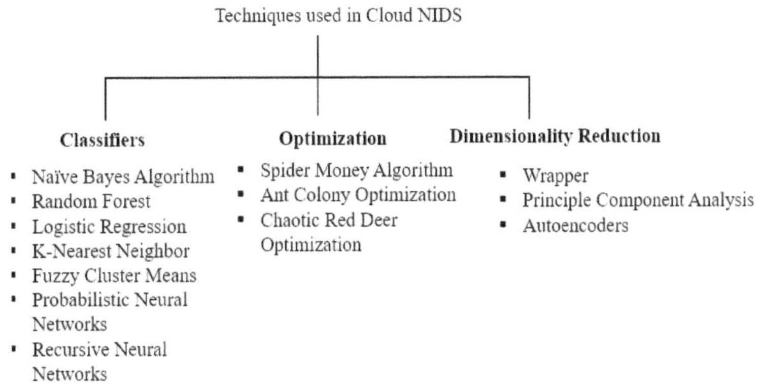

Fig.5.1. Algorithms used in Developing NIDS for Cloud Computing

techniques include the classifiers as well as dimensional reduction techniques. We have also analyzed the performance of these classifiers and came up with the classifier that has produced the best confusion matrix results. We have ended the chapter by providing the concluding remarks.

5.2. Classification Algorithms

Classifiers were used either for the classification of the data or in some cases they were used in wrapper methods for feature selection.

5.2.1. Naïve Bayes Algorithm

Naive Bayes algorithm is a statistical technique that relies on the theory of probability. It uses conditional probability to predict whether the network traffic is normal, or intrusion based on some already known information (training). Naive Bayes algorithm is expressed as in equation 5.1 [105],[91].

$$P(C_L \mid F_1,\ldots,F_n) = \frac{1}{Z} p(C_L) \prod_{i=1}^{n} p(F_i \mid C_L) \qquad \text{(5.1)}$$

This essentially means that the probability of level L for a already given class C, when some of the facts are known through features ($F_1 \ldots F_n$) is the product of the conditional probabilities of individual facts or evidence. 1/Z is the scaling factor of probability [57].

5.2.2. Radom Forest

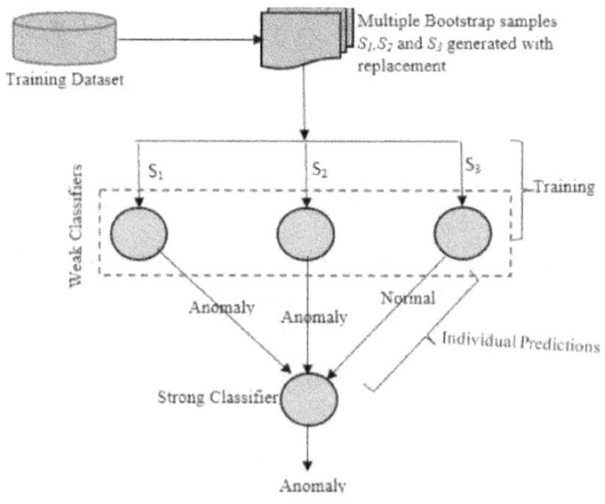

Fig. 5.2 Random Forest Model

Random forest is an ensemble supervised machine learning algorithm that aggregates the predictions from multiple trees to improve the prediction. In this model, multiple training datasets are created from the original sample using bootstrap sampling. These datasets are then used to train multiple but same machine learning model. Prediction from all these models are the aggregated and decision is taken using voting method [35]. Figure 5.2 shows a random forest model with three decision trees and a voting mechanism to make the decision.

5.2.3. Logistic regression

Regression is mainly the technique that establishes the relationship between one dependent variable and one or more independent variable. The linear regression is mainly concerned with the numeric variables; however, the logistic regression can be used to develop a model for categorical variables as well as. In general, this technique finds a line that separates two classes of variables. An equation with three independent variables x, y, and z and one dependent variable Y is given as equation 5.2 [106].

$$Y = ax+by+cz \qquad \text{.......................} (5.2)$$

5.2.4. Clustering Methods

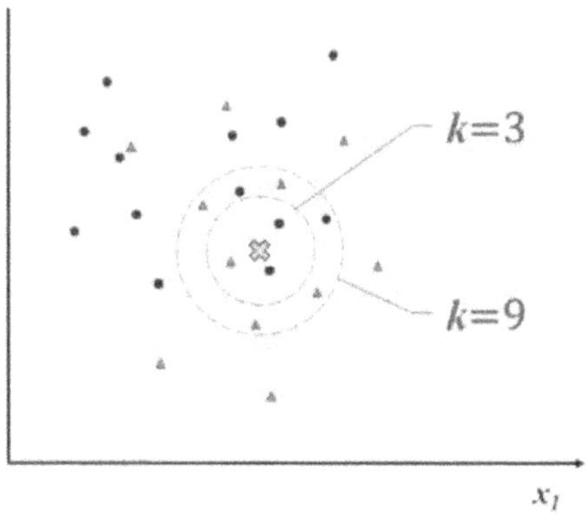

Fig. 5.3. Cluster boundaries with k=5 and 9

k-Means clustering is an unsupervised machine learning approach. The principle of such models is that given the training data set with *Xi* features and *Ci* classes and the test dataset with the same features as the training dataset, k-NN identifies each record in the test dataset that is most “similar” to k records in the training dataset. Similarity is measured in terms of distance between features like Euclidean distance [27]. Figure 5.3 shows the cluster boundaries for outliers and normal data points for k=3 and 9. k-means clustering has been effectively used in combination with Snort to detect the anomalies in the data in the cloud environment [67].

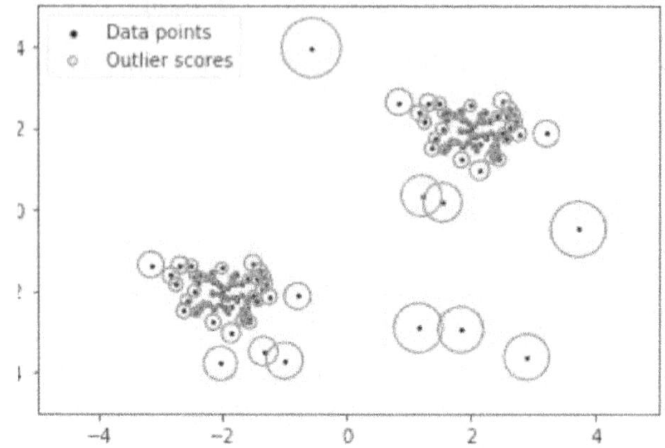

Fig.5.4. Density Based Clusters

Clusters used for classification can be population based or density based. Local Outlier Factor (LOF) is a density-based clustering techniques which detects examples with less neighbor density as outliers. Figure 5.4 shows the clusters with high density and low density. The data points with the low density are treated as outliers. This method has been used to find the anomalies in cloud virtual machines using the network data [107].

5.2.5. Neural Networks

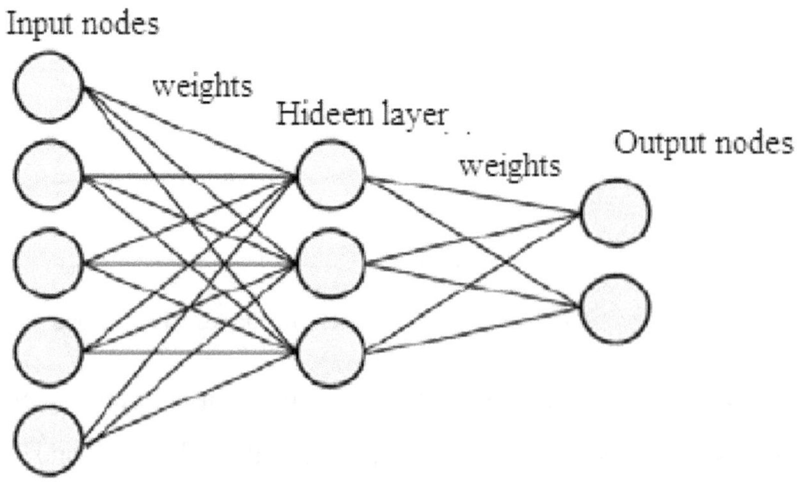

Fig 5.5. Neural Network with hidden Layer

ANN models the relationship between the input variables and the output. To create the model, it uses input nodes, multiple hidden layers of processing nodes and one or more output nodes as shown in Figure 5.5. Output of ANN is given in equation 5.3.

$$y(x) = f\left(\sum_{i=1}^{n} w_i x_i\right) \quad \text{.......................... 5.3}$$

Where y is the output of the neural network model, *wi* are the weights and *xi* is the set of inputs.

In the recent literatures, ANN has been used as the main classifier to detect anomalies [38].[108]. However, ANN is slow to train and computationally expensive, especially with more hidden layers. Also, the fact that ANN requires a lot of training data to give good results, there have been some challenges in implementing this model like availability of adequate training dataset etc.

Recurrent Neural networks are a variation of ANN that is used to process the sequential data. Like the neural networks which used backpropagation to improve results, RNN uses Backpropagation through Time (BPTT) algorithm to produce better output [93].

Probabilistic neural networks use probability theory while doing the classification. It is same as traditional ANNs but do not use backpropagation like ANN. Thus, it is less computationally expensive that ANNs [87].

5.2.6. Support Vector Machines

SVMs combine the features of linear regression as well as clustering. The model creates a linear line called the boundary that partitions the elements such that the elements on one partition are homogeneous and the elements between two partitions are heterogeneous. Figure 5.6 shows an SVM that creates two classes based on two shapes.

SVM has been used in combination with FCM to detect the anomalies [109]. FCM is used to split the data into finite clusters to improve the efficiency of SVM. Each cluster is used to train SVM units, and the result is then aggregated using Fuzzy aggregation model. Using all the features in the data could slow down the detection process in the cloud. Reducing the dimensionality could improve the process of detection. Convolutional Neural networks have also been used in combination with SVM to detect the type of attacks [110]. SVM has also been used as a stand-alone model for classification by checking the variation in the speed of traffic.

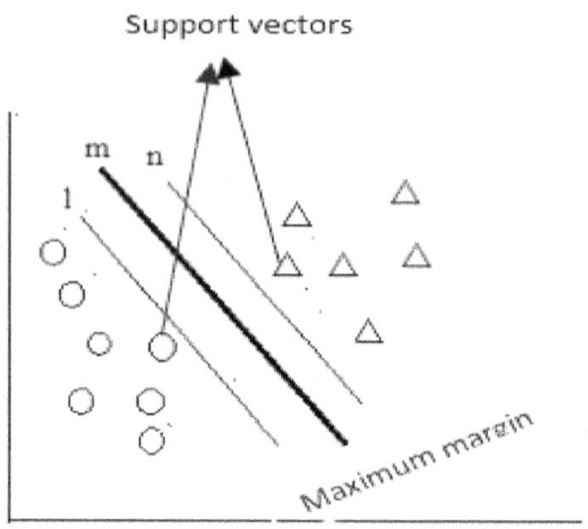

Fig.5.6. Portioning Data using Support Vector Machine using Hyperplane.

5..3. Dimensional Reduction Techniques

Dimensional reduction techniques were used to reduce the number of features in the dataset and increase the accuracy of classification results.

5.3.1 Wrapper Method

Wrapper methods evaluate a set of attributes on a classifier to assess its predictive performance. Wrapper methods are considered to be more efficient as they take into consideration the classifier hypothesis [46]. While filter method is independent of any classifier, wrapper method may give different results when different classifiers are used [24]. Wrapper methods uses greedy search to search the feature space and select the features for the evaluation.

5.3.2. Principal Component Analysis

Since the data in the cloud is high dimensional in nature, PCA is used to transform the data from high feature space to low feature space. It removes the collinearity in the variables and transforms original set of creates a new set of variables. PCA is a multivariate method that extracts the most important features from the set of interrelated features that have high impact on the dependent variable. The newly created features are called as Principal components [26].

5.3.3 Autoencoders

Autoencoders are unsupervised neural networks that can be used to encode or compress the original data into a transformed dataset and then reconstruct the original data from the transformed dataset [111]. The two components of the autoencoder are encoder and decoder. Encoder reduces the dimensions of the input dataset and hence compresses the data. Decoder, on the other rebuilds the original input from the transformed output. Architecture of autoencoder is similar to that of an ANN as shown in figure 5.7. However, the final hidden layer of autoencoder which is also called the bottleneck layers is responsible for transforming the encoded data back to the original representation [112].

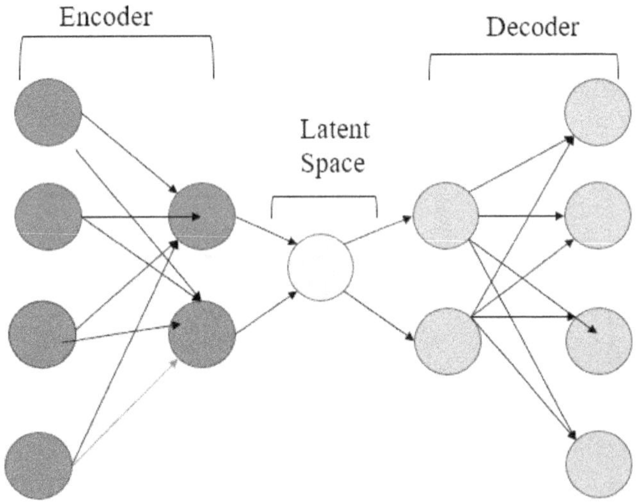

Fig.5.7 Autoencoder with two encoder and two decoder layers

5.4. Optimization Methods

Optimization methods are used to optimize the solution. They have been successfully used to optimize the feature search to select the most relevant features.

5.4.1 Spider Monkey Algorithm

Spider Monkey Algorithm (SMA) is a meta-heuristic algorithm which is inspired by social behavior of spider-monkeys. It is an optimization algorithm that mimics the behavior of spider monkeys during searching of food. SMA works in multiple phases. During the initial phase a population of N spider monkeys is generated. During the local leader phase, spider monkeys update their position using the greedy search if the new position is better than the present in terms of individual fitness. During the global leader phase, global fitness of the whole swarm is found. During feature selection SMA can be used to optimize the feature space to select most relevant features [113].

5.4.2. Ant Colony Optimization

Ant Colony Optimization is used to optimize as solution to a problem and is inspired by the foraging behavior of ants. The main concept of this algorithm is based on the movements of ants when they are in search of food. Figure 5.8 shows the algorithm for Ant Colony Optimization.

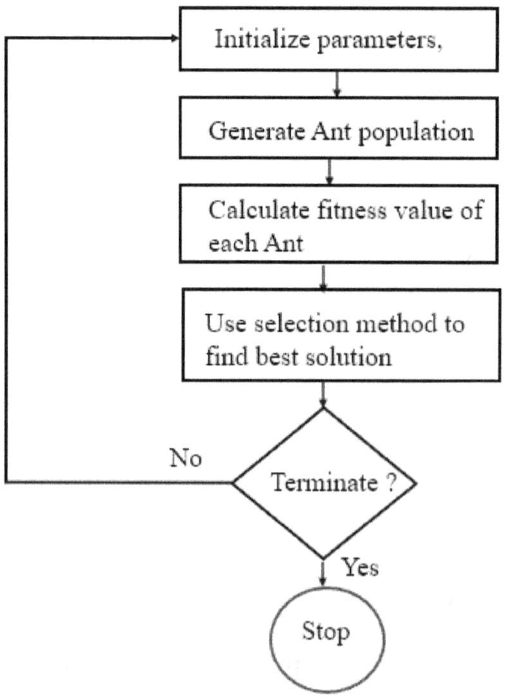

Fig.5.8. Ant Colony Optimization Algorithm

5.5. Performance Comparison

Table 5.1 summarizes the recent techniques that have been used in developing the ANIDS models and the performance of these models. During review of the recent techniques used in cloud computing, it was observed that most of the time ANN methods were used in combination with other dimensional reduction techniques to develop the ANIDS for cloud computing.

Table 5.1. Performance of Classifiers

Performance Metrics	Hybrid KNN	Hybrid FCN	Hybrid ANN	Hybrid SVM
Avg. Accuracy	0.86	0.86	0.98	0.95
Avg. TPR	0.98	0.88	0.95	0.86
Avg. FPR	0.10	0.05	0.08	0.03

Figure 5.9 shows the average accuracy of the of the classifiers in the best case. In other words, the graph shows the best average accuracy that was produced by each classifier. It can be seen that in terms of accuracy ANN has given best results as compared to other classifiers. The reason is that ANN can be tuned to solve complex problems by increasing the number of hidden layers or nodes. This can be done to increase the accuracy of classifier to some extent.

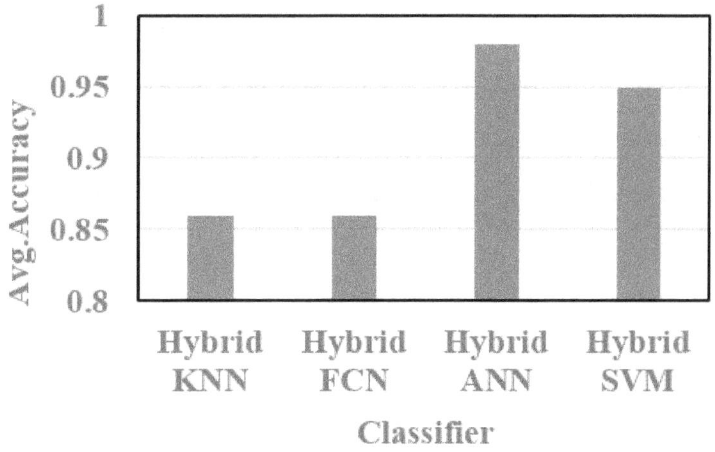

Fig. 5.9. Performance on Accuracy

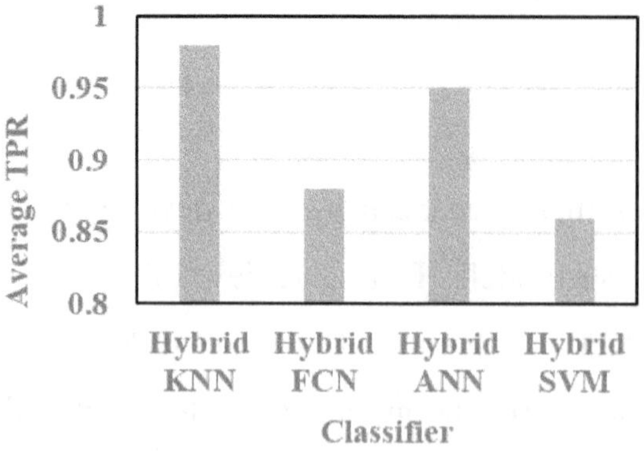

Fig. 5.10. Performance on TPR

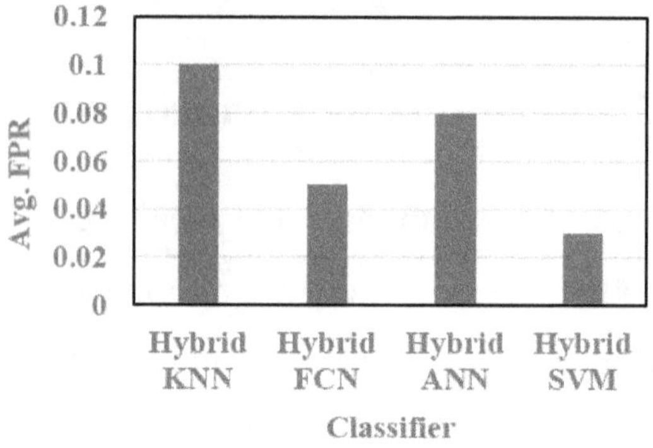

Fig. 5.11. Performance on FPR

Figure 5.10 shows the best average performance of classifiers on TPR. It is clear from this figure that kNN has performed slightly better than other classifiers in detecting true positive alarms. However, from the figure 5.11 we see that ANN has performed much better in detecting FPR than other classifiers. Looking at these three graphs together, we can clearly see that ANN has been able to produce better classification results than other classifiers during recent years.

5.6. Conclusion

In this chapter we have briefly explored various techniques that were used in developing the Anomaly Based Network Intrusion Detection System for cloud environment. We observed that neural network technique have been used most of the time. Main observation was that neural networks have been able to give better results as compared to other machine learning techniques. This can be due to the fact neural networks can solve complex problems by increasing the number of layers or number of nodes per layer. However, in such cases the model will be computationally expensive.

Chapter 6

CHAPTER 6 : PROPOSED COLLABORATIVE BASED NETWORK INTRUSION DETECTION SYSTEM

6.1. Introduction

Cloud computing has emerged as a popular information technology platform for various organizations as well as home users. It has helped the organizations to reduce the operational cost by providing on-demand and scalable resources using pay-as-you-go model. This means that customers need to pay only for the services that they use. Hence the charges are metered. Since the cloud is accessed using the public network, it presents some security challenges. The openness of the cloud, large attack surface area and the shared resources are some of the reasons that attackers find a way to intrude into the cloud environment and steal or damage the cloud resources which include the critical data of end users. NIDS is a tool that can be used to detect the intrusions in the network and prevent or at least minimize the damages done by these intruders. Since the cloud is characterized by high volume of inbound network traffic, which is high dimensional in nature, NIDS needs to detect intrusions by examining fewer features of these network packets [114]. This will speed up the detection rate. In this chapter, we have proposed a Collaborative Network Intrusion Detection System that used deep learning and gradient boosted tree as the first and second-layer classifiers, which were optimized using evolutionary algorithm. correlation based feature selection method was used to select most relevant features and was optimized using PSO. To extract features, we used SVD method which was also optimized using evolutionary algorithm. We trained and tested our model on NSL-KDD dataset and achieved an improved confusion matrix results as compared to earlier proposed models. We achieved better performance using only seven features which was one of the important objectives of our work. Proposed model can be used in cloud computing for intrusion detection as it uses fewer features to make the predictions with better accuracy as compared to earlier models.

6.2. Proposed Optimized Collaborative Network Intrusion Detection System

Architecture of the of the proposed model is shown in the figure 6.1. This model was developed in different phase as shown in figure 6.2.

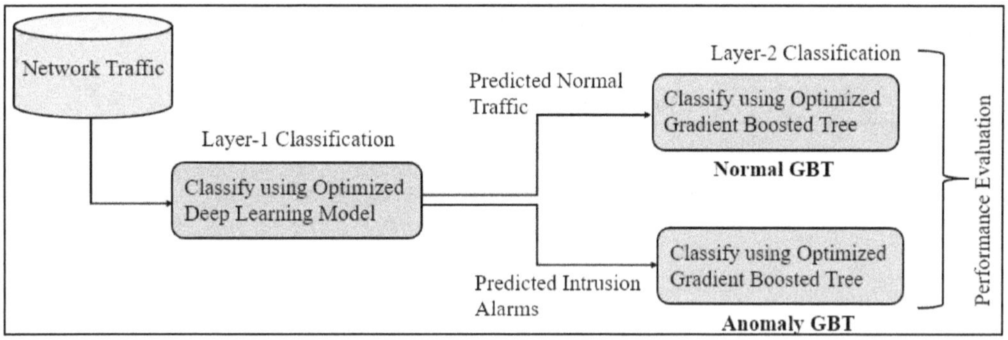

Fig.6.1. Architecture of Proposed Model

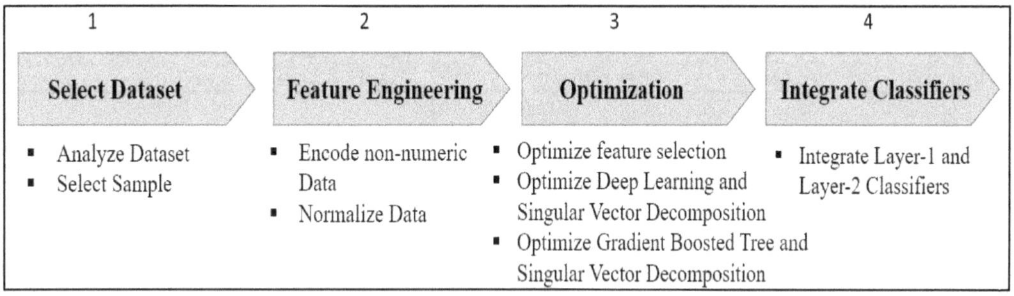

Fig.6.2. Different Phase in Development of Proposed Model

6.2.1 Dataset Selection

Network data is an essential component in developing and evaluating a NIDS. Some datasets that have been used for this purpose include NSL-KDD and CIDDS dataset. NSL-KDD dataset does not contain any redundant or duplicate examples which prevents biased result towards more frequently occurring examples. This property of NSL-KDD dataset makes it more popular among the researchers. Hence, we also selected this dataset for our work. NSL-KDD dataset has been used by the many researchers in developing and evaluating the NIDS models [30]. However, before developing the proposed model, this data set needed to be processed using feature engineering techniques.

6.2.2. Dataset Analysis

NSL-KDD dataset contains 125973 examples including 67343 examples which are labelled as normal and 58630 examples which are labelled as anomaly. Hence the dataset is well balanced. There are 42 features in the dataset which include one class feature. NSL-KDD dataset contains the nominal, polynomial and numeric features. Properties of NSL-KDD are shown in the table 6.1.

Table 6.1. Properties of NSL-KDD Dataset

Feature	Type	Feature	Type	Feature	Type
duration	Integer	su_attempted	Integer	same_srv_rate	Real
protocol_type	Polynomial	num_root	Integer	diff_srv_rate	Real
service	Polynomial	num_file_creations	Integer	srv_diff_host_rate	Real
flag	Polynomial	num_shells	Integer	dst_host_count	Integer
src_bytes	Integer	num_access_files	Integer	dst_host_srv_count	Integer
dst_bytes	Integer	num_outbound_cmds	Integer	dst_host_same_srv_rate	Real
land	Integer	is_host_login	Integer	dst_host_diff_srv_rate	Real
wrong_fragment	Integer	is_guest_login	Integer	dst_host_same_src_port_rate	Real
urgent	Integer	count	Integer	dst_host_srv_diff_host_rate	Real
hot	Integer	srv_count	Integer	dst_host_serror_rate	Real
num_failed_logins	Integer	serror_rate	Real	dst_host_srv_serror_rate	Real
logged_in	Integer	srv_serror_rate	Real	dst_host_rerror_rate	Real
num_compromised	Integer	rerror_rate	Real	dst_host_srv_rerror_rate	Real
root_shell	Integer	srv_rerror_rate	Real	class	Polynomial

6.2.3. Sample Selection for Proposed Model

Out of 125973 examples, we randomly selected 9448 examples for our work. Our sample included 4410 examples which were labelled as anomalies and 5028 examples were labelled as normal. Hence sample dataset was well balanced. In the sample dataset, 70% of the examples were used for training purpose and 30% were used for testing purpose.

6.2.4 Feature Engineering Process

Feature engineering includes a set of activities that can transform the existing features into a new set of features, encode the features, select the relevant features, or re-scale the features. Feature engineering helps to make dataset features compatible with machine l\earning algorithm or reduce the dimensionality of the dataset. In our work we have used various feature engineering techniques to manipulate the features to make the model less complex and more efficient [88].

a) Feature Encoding

Since our dataset consisted of some nominal and polynomial features, we needed to transform these features to numeric type. Table 6.2 shows the description of nominal and numeric features. Hence, prior to normalization, nominal and polynomial features were encoded to the numeric values and binomial features were encoded to 0 and 1. Thus we had a transformed dataset with 41 numeric features and one class variable.

Table 6.2. Encoding Features of NSL-KDD dataset

Feature Type	Number of features	Encoding needed
Numeric	34	X
Nominal	8	√

b) Normalization

One of the common problems with numeric features is that some features may have a larger scale than others. This means that some features may have larger numeric values than others. In such a case the machine learning algorithm tends to bias the results towards the features with high numeric values [83]. To overcome this problem, we adjusted all the features using a common scale. This means that all the numeric values were scaled between two values. Common algorithm that are used for this purpose are minimax algorithm and Z-transformation. Prior to feature selection process, features were normalized using minimax algorithm to scale all the numeric values between 1-10. This was done because scaling the values between 0 to 1 resulted in small duplicate numbers. So, we had to use a larger scale. Formula for minimax algorithm is shown in equation 6.1.

$$z = \frac{x - x_i}{x_j - x_i} \qquad \text{............} \quad (6.1)$$

where,

z is the transformed value, x is the value to be scaled, x_i is the minimum value of x and x_j is the maximum value of x. After transformation, all the numeric values lied within a range of 1 to 10.

c) Collinearity Between Independent Features

Cloud computing is characterized by high volume of incoming data which is high dimensional in nature [16]. Thus, to decrease the detection time, NIDS should be able to make prediction by inspecting fewer features but with greater accuracy. For this purpose, we needed to select the most relevant features and discard those features which had less or no impact on class variable. Thus, the features that constituted the noise needed to be identified and removed and features with high collinearity also needed to be identified. So, we used the correlation matrix to identify the features with high collinearity.

As can be seen from the correlation matrix in figure 6.3, there was a strong collinearity between some of the independent features which could impact the prediction. Some independent features were also having little or no impact on the class variable. Hence, we had to identify and remove the irrelevant features and then remove collinearity between the independent features. This was done by using optimized feature selection followed by optimized feature extraction.

-0.079	0.156	0.205	0.179	-0.106	0.170	-0.060	-0.144	-0.144	-0.094	-0.112	0.003
-0.229	0.397	-0.559	-0.622	-0.014	-0.271	-0.171	0.979	0.983	-0.234	-0.229	0.647
-0.227	0.392	-0.554	-0.616	-0.028	-0.270	-0.168	0.978	0.986	-0.238	-0.234	0.642
-0.069	0.069	-0.253	-0.260	0.371	0.035	0.039	-0.228	-0.229	0.928	0.969	0.258
-0.072	0.077	-0.255	-0.263	0.382	0.032	0.033	-0.228	-0.232	0.918	0.974	0.262
0.282	-0.518	0.705	0.790	-0.157	0.283	0.219	-0.758	-0.761	-0.216	-0.216	-0.748
-0.110	0.194	-0.298	-0.331	0.478	-0.024	-0.085	0.055	0.048	0.223	0.282	0.205
1	-0.364	0.146	0.294	-0.120	0.247	0.384	-0.233	-0.231	-0.072	-0.077	-0.120
-0.364	1	-0.302	-0.518	0.131	-0.305	-0.453	0.414	0.408	0.079	0.074	0.375
0.146	-0.302	1	0.895	-0.390	-0.072	-0.013	-0.572	-0.567	-0.255	-0.261	-0.728
0.294	-0.518	0.895	1	-0.419	0.131	0.197	-0.638	-0.630	-0.263	-0.267	-0.693
-0.120	0.131	-0.390	-0.419	1	0.230	0.034	-0.013	-0.024	0.406	0.385	0.245

Fig.6.3. Correlation Matrix of NSL-KDD dataset features

d) Optimized Feature Selection using Correlation and Particle Swarm Optimization

Correlation Based Feature Selection (CfsSubseteval) measures the inter-correlation in addition to the correlation with the class variable while selecting the features. This makes it a preferred choice for feature selection method. This method uses Person's correlation coefficient as measure to select the relevant features for classification [97],[96]. Person correlation coefficient is given by equation 6.2

$$r_{xy} = \frac{\sum_{i=1}^{n}(x_i - \bar{x})(y_i - \bar{y})}{\sqrt{\sum_{i=1}^{n}(x_i - \bar{x})^2}\sqrt{\sum_{i=1}^{n}(y_i - \bar{y})^2}} \quad \text{................ (6.2)}$$

During the feature selection process, PSO was used to optimize the search of features in the features space [87]. In case of features selection, features are considered as the particles in the feature space. Each particle evaluates the objective function at its current location and decides the next movement [100]. The next movement is decided based on fitness value (accuracy) of next location and the current location. This process continues and the flock or swarm (features) moves close to the solution or the optimum value of fitness function. In our work, fitness function is the accuracy function that needs to be maximized and the features act as the swarm that work towards getting the better accuracy.

We used correlation-based feature selection method with 20-fold cross validation to evaluate the accuracy of features on deep learning model. PSO algorithm is shown in Algorithm 6.1, where $\bar{x}$ is the current position of the particle in D dimensional search space, p_i is the previous best position,mv_i is the velocity, pop_t stores the optimized result after each iteration. Table 6.3 shows the features that were selected by feature selection algorithm. There were some features which showed number of folds as 0%. This means that these features were not considered relevant and were never used in the evaluation process by correlation algorithm. Hence, we excluded such features from our dataset and considered only those features which showed number of folds as greater than zero. Hence in the new transformed dataset, we now had 26 features including the class feature.

The process of optimized feature selection helped to select the most relevant 26 features which could give the best accuracy on the classifier. However, there were still some features which could have collinearity and impact the accuracy of the model. So, our next task was to combine those features which had high collinearity using feature extraction techniques. So, feature selection method was followed by optimized feature extraction method [115]. We optimized SVD on both, DNN and GBT.

Algorithm 6.1. Particle Swarm Optimization Algorithm

Initialize the population in search space
for each particle find the desired optimum fitness function.
start loop
compare the fitness of particle with $p_{opt.}$
If the current value is better than p_{opt}
start loop
set p_{opt} equal to current value
set p_i to the current location $\bar{x}$
end loop.
identify the most optimized particle in neighborhood and assign its index to variable g.
modify the velocity and position of the particle using equation.

$$\begin{cases} v_i \leftarrow v_i + U(0, \phi_1) \otimes (p_i - x_i) + U(0, \phi_2) \otimes (p_g - x_i), \\ x_i \leftarrow x_i + v_i. \end{cases}$$

if criteria is met exit the loop
end loop.

Table 6.3. Features Selected using Feature Selection Method

duration	num_file_creations	num_compromised	dst_host_diff_srv_rate
protocol_type	num_shells	su_attempted	dst_host_same_src_port_rate
service	num_access_files	num_root	dst_host_rerror_rate
dst_bytes	num_outbound_cmds	srv_count	dst_host_count
land	is_host_login	srv_rerror_rate	class
wrong_fragment	is_guest_login	same_srv_rate	
logged_in	count	diff_srv_rate	

e) Optimizing Deep Learning and Singular Vector Decomposition Parameters (Layer-1 Optimization)

SVD is based on a principle that a matrix can be broken down into three different matrices. These include an Orthogonal matrix U, a diagonal matrix S and transpose of orthogonal matrix V. If H is a matrix with dimensions (m,n) then SVD is given by equation 6.3 [101].

SVD can be used for feature extraction and performs well as compared to self-organizing maps [101]. Due to the performance of SVD, we used it as the feature extraction technique.

$$H = USV^T \quad \text{.....................} (6.3)$$

Where $U^TU = V^TV$

Subject to condition $S = diag(a_1 .... a_n)$

Where $a_i > 0$ *for* $1 \leq i \leq r$ *and* $a_j = 0$ *for* $j \geq r+1$

DNN is a type of ANN which uses multiple hidden layers to solve the complex problems. Each layers has two or more than two nodes to perform the computation. DNN accepts the input through input nodes and produces the output through output nodes [106]. A typical DNN with two input nodes is shown in the figure 6.4.

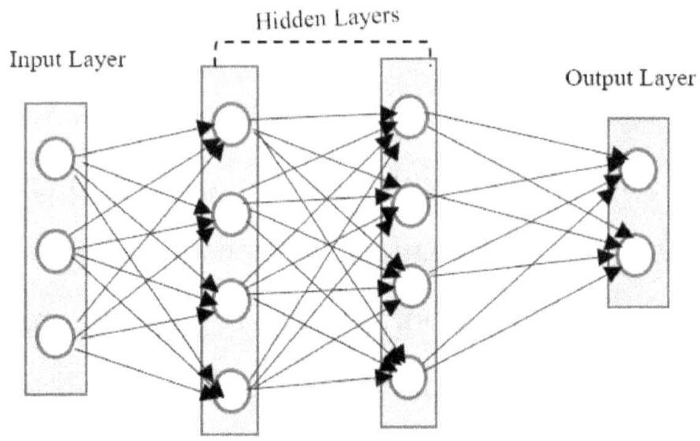

Fig. 6.4. Deep Learning Network with two Hidden Layers

Since, in cloud computing NIDS should be able to make the predictions by processing fewer features, we needed to optimize the dimensions of the dataset prior to training and testing of our proposed model. Hence, during optimization phase, the feature extraction technique as well as DNN hyperparameters were optimized and recoded to get the best accuracy on DNN classifier. For SVD, number of dimensions was optimized and for DNN, number of epochs and epsilon were optimized. Genetic algorithm was used to optimize both SVD and DNN hyperparameters. In our work we used tanh as the activation function as it gives better results. Backpropagation with gradient descent method was used to adjust the weights [27]. Final prediction was received through the output nodes. Thus, DNN predicted the examples as "normal" and "anomaly". DNN can be thought of a series of functions that applies the transformation function at each hidden layer and produces the output through the output nodes ,

Genetic algorithm belongs to evolutionary algorithm class. Evolutionary algorithm is a term that us used to describe the algorithms that use Darwin's evolutionary theory to solve complex problems [104]. One such algorithm is genetic algorithm. Genetic algorithm is an adaptive, heuristics algorithm based in genetics and natural selection process that is used to generate high quality solutions for optimization problem [103]. Genetic algorithms search mimic the process of natural evolution to optimize the results. In our experiment, we used the genetic algorithm that uses the different evolutionary phases like Initialize, Mutation, Crossover, Evaluate and Selection [104]. Various steps that genetic algorithm used to select the optimized hyperparameters include.

Initialize Population and Crossover

We needed to select the initial population for crossover. During Crossover, the selected population started to select the features, dimensions, epsilon and epochs and evaluate the fitness value (Accuracy). Poor performing hyperparameters were dropped and replaced new hyperparameters. This process continued till we got the best accuracy. We selected the initial population as 5. We used the crossover probability of 0.5% and we stopped the algorithm when there was no improvement in 5 consecutive iterations [89].

Mutation

Mutation was performed by flipping the hyperparameters i.e, by setting the used hyperparameters to unused. We configured the mutation type as Gaussian and selection type as tournament. Mutation allows to slightly change the parent information while slightly adding the new child information.

Evaluate

In this step, all the hyperparameters generated during crossover and mutation were evaluated iteratively using DNN. The accuracy for each set of hyperparameters was calculated. Hyperparameters that produced higher accuracy were more likely to be selected for next generation. This process continued till we didn't have any improvement in accuracy for 5 consecutive iterations. At this point the algorithm stopped, and the best performing hyperparameters were selected.

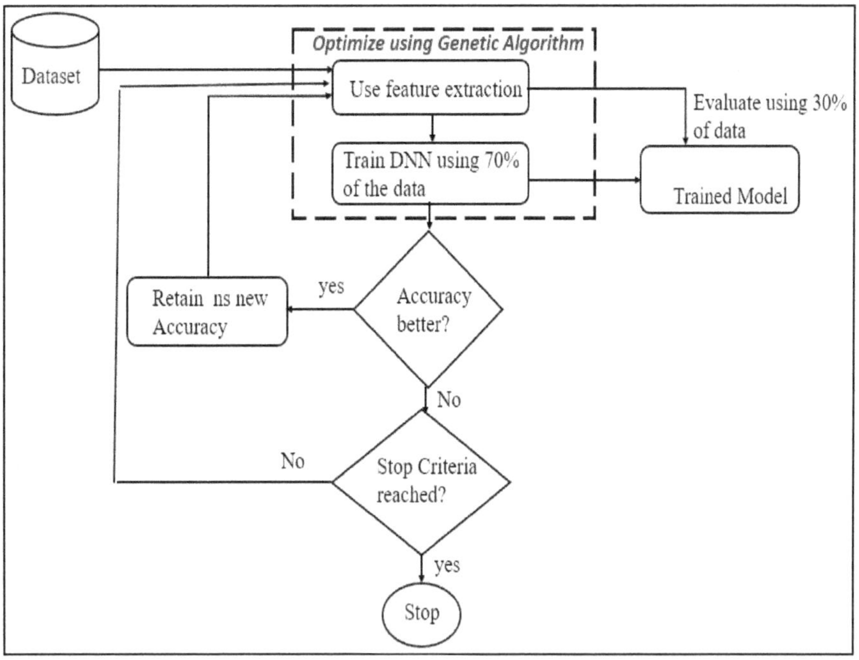

Fig.6.5. Optimization Process of SVD and Deep Learning Model

Selection

In this process, a selection mechanism was used to select the best performing hyperparameters for the next generation. In our work, we used tournament selection method. Tournament method selects the features with the best accuracy for the next generation [116].

To optimize the SVD and DNN, we used 70% of the data as training dataset and rest 30% of the data as the test dataset. The optimization process is shown in figure 6.5 and the output of the optimization process is shown in the table 6.4. It is clear from this figure that we were able to achieve an average accuracy of 0.909 by using only 6 dimensions of the dataset. The optimized values of epochs and epsilon are also shown in the table. Hence, we treated these values as the optimized configuration for layer-1 of the NIDS model.

Table 6.4. Optimized Hyperparameters of 50x50 Deep Learning Network

Seed	Accuracy	Epochs	Epsilon	Dimensions
1	0.900	15.20165343	21.91736797	4
2	0.909	8.971967825	69.77123432	6
3	0.904	15.19921656	21.9245967	5
4	0.913	15.17845983	21.95823211	4
5	0.909	20.79009211	39.29182396	11
Average	**0.909**	**15.06827795**	**34.97265102**	**6**

f) Optimizing Singular Vector Decomposition and Gradient Boosted Tree (Layer-2 Optimization)

Like Layer-1 optimization, we optimized the Layer-2 of proposed model using genetic algorithm. We simultaneously optimized SVD to select the least features and Gradient Boosted Tree (GBT) to evaluate the accuracy of the predictions. Our aim was to extract least number of features that could produce best accuracy on GBT.

GBT is a sequential arrangement of decision trees that used to solve complex problems [91]. Each new tree in the arrangement tries to reduce the errors made by the previous tree. Thus, each new tree is an improvement over the previous tree and produces better accuracy. In our work, we optimized the number of trees along with the depth of GBT. The output of SVD and GBT hyperparameter optimization are shown in table 6.5. we clearly see that we achieved an accuracy of 0.983 using 7 dimensions only.

Table 6.5. Optimized Hyperparameters of Gradient Boosted Tree

Seed	Features	Depth	Trees	Accuracy
1	9	13	147	0.980
2	6	67	210	0.987
3	6	67	210	0.987
4	11	66	118	0.980
5	**6**	67	210	0.981
Average	**7.6**	**56**	**179**	**0.983**

6.3. Integrating Layer-1 and Layer-2 of Proposed Model

The optimized Layer-1 and Layer-2 classifiers were connected as shown in figure 6.1. DNN and GBT were configured with the optimized parameters and SVD was configured to extract 7 (highest of two results) features. We used 70% of the data for the training purposes and 30% data for testing purposes. Working algorithm for the proposed model is shown in the Algorithm 6.2. Initially, DNN classified the network traffic into normal and intrusion packets. Normal packets were sent to "*Normal GBT*" and anomaly packets were sent to "*Anomaly GBT*" for further classification. These GBTs classified normal and anomaly traffic further into TPs, TNs, FPs and FNs Thus, the normal and anomaly packets as predicted by DNN were further classified separately by Normal and Anomaly GBTs.

Algorithm 6.2. Working Algorithm of Proposed Model

Input : 70% of training data and 30 % test data
Output: Intrusion Alarms
// Feature Selection and Optimization
   Select features using Optimized feature selection method
   Optimize SVD on Deep learning model
   Optimize SVD on gradient Boosted Tree.
// Layer-1 Classification
   Classify test data into normal and anomaly using deep learning
   Filter normal and anomaly traffic
//Layer-2 Classification
   Classify normal data into True normal and False positive using Normal GBT
   Classify anomaly data into True anomaly and false positives using Anomaly GBT
// Alarm
   Raise positive alarms

6.4. Experimental Setup

Experiment was performed on windows10 machine with 16GB RAM and Intel i5 core. RapidMiner Studio and Weka machine learning tool were used for the simulations. Weka was used for feature selection and rest of the work was done on RapidMiner. Layout of the actual process used for simulating the proposed NIDS model in RapidMiner is shown in figure 6.6

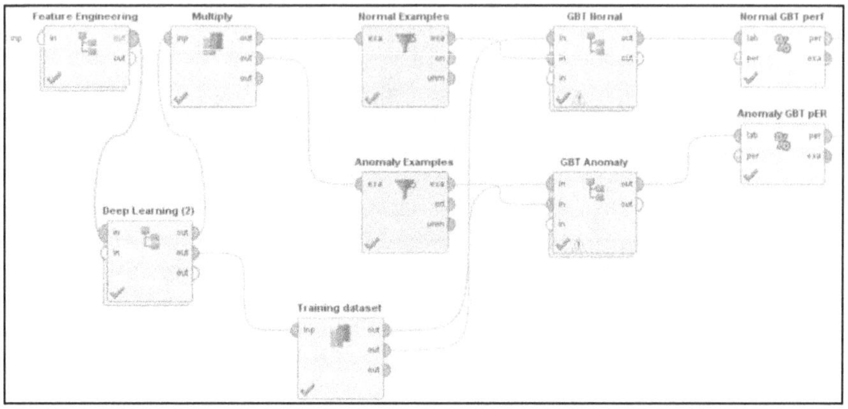

Fig. 6.6. Process Layout of Proposed Model using RapidMiner

Figures 6.7, 6.8 and 6.9 shows the RapidMiner process layout for feature engineering process, deep learning process and GBT process for classification.

While performing the experiment, we took 5 observations for the classifications with 5 different seed values to increase the reliability of the result. The average value of the observation was considered as the final value. These include the observation during the optimization process as well as intrusion detection process.

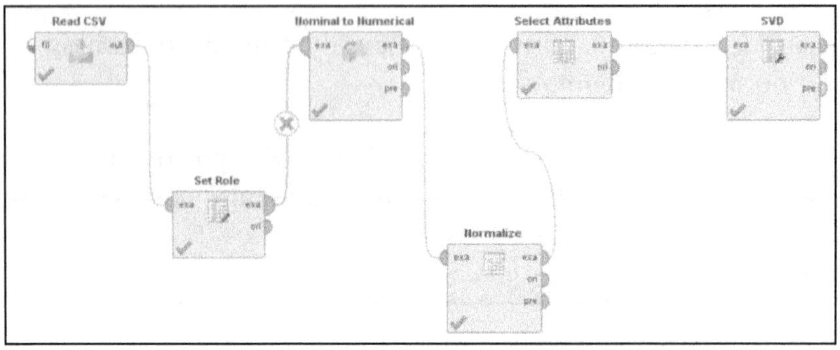

Fig. 6.7. Feature Engineering process

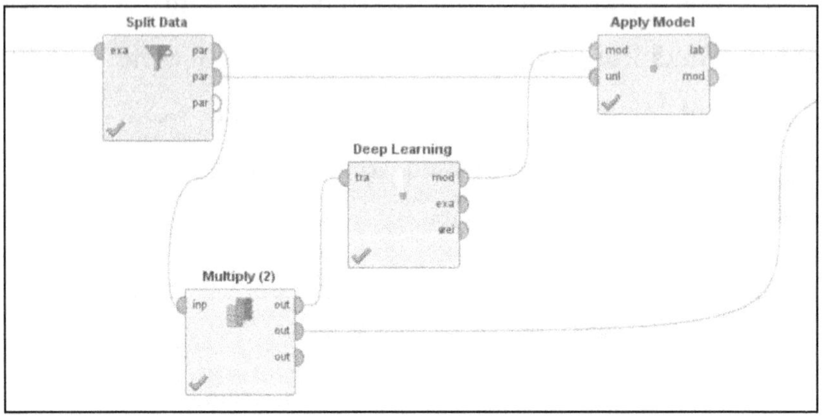

Fig. 6.8. Deep Learning Classification Process

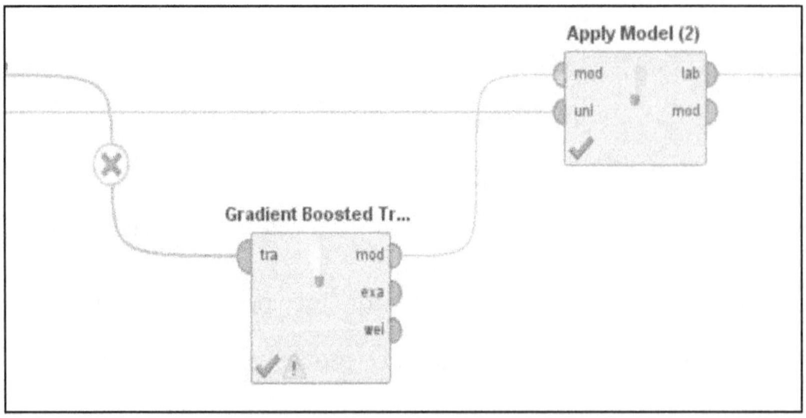

Fig. 6.9. Gradient Boosted Tree Classification Process

6.5. Results and Discussion

Table 6.6 shows the classification results of the normal traffic by *Normal GBT*, since *Normal GBT* was fed with only the normal examples from DNN. The experiment was repeated multiple times with different seed values to get reliable results. As can be seen from the table, most of the true normal packets were detected which cloud be forwarded to the users. False Negatives refer to those packets which were intrusions but were treated as normal packets by proposed model. Our proposed model was able to minimize these alarms which can effectively prevent or minimize intrusions.

Table 6. 6. Classification of normal examples using Normal GBT

Seed	Accuracy	True Positives	True Negatives	False Positives	False Negatives
1	1.00	6	3	0	0
2	0.97	432	1426	28	26
3	0.97	460	1411	21	36
4	0.97	290	1368	24	35
5	0.97	292	1292	23	28

Table 6.7 shows the classification of anomaly alarms into TPs, TNs, PFs and FNs by *Anomaly GBT*. We can clearly see from the table that most of the TP alarms have been detected and raised by the proposed NIDS model. All these alarms were the real intrusion that could impact the cloud. FNs refer to the packets that were treated as normal packets by our model, but they were intrusions. Proposed model has been able to minimize these FNs which means that most of the real intrusion packets have been converted into the alarms.

Table 6.7. Classification of Anomaly Traffic using Anomaly GBT

Seed	Accuracy	True Positives	True Negatives	False Positives	False Negatives
1	0.97	1296	1458	31	40
2	0.99	859	50	11	2
3	0.98	835	56	8	7
4	0.99	967	139	6	5
5	0.99	1027	163	6	3

Table 6.8 shows the combined performance of all the components of the proposed model. It can be clearly seen that we were able to achieve an average accuracy of 0.98 which means that we were able to detect the true alarms with 98% accuracy. True alarms being close to 1 means that we been able to detect most of the true alarms while minimizing the false alarms and keeping them close to 0.

Table 6.8. Performance of the Proposed Model

Seed	True Positive Rate	True Negative Rate	False Positive Rate	False Negative Rate	Accuracy
1	0.986	0.979	0.021	0.030	0.97
2	0.990	0.974	0.026	0.021	0.98
3	0.985	0.981	0.019	0.032	0.97
4	0.986	0.980	0.020	0.031	0.98
5	0.989	0.980	0.020	0.023	0.98
Average	**0.987**	**0.979**	**0.021**	**0.030**	**0.98**

Table 6.9 shows the comparison of proposed model with NIDS models that were proposed in earlier literature. It can be seen from this table that the number of features used for the classification was too high. In fact, least number of features in these literatures was 15. Our proposed model used only 7 features for the classification which can reduce the detection time. We have also achieved an improved accuracy of 0.98 as compared to earlier models which is shown in figure 6.10. More importantly FPR has been minimized as compared to earlier models which is an improvement over these models. This can be seen from figure 6.12. As shown in figure 6.11, we have been able to increase the TPR which means that we were able to detect more intrusions as compared to other models.

From the analysis of the confusion matrix parameters, we observe that the proposed model has better results in comparison two earlier models that were proposed for cloud security. The proposed model has higher accuracy which means that it will be able to able to detect more true alarms as compared to earlier proposed models/ This model also has lower false positives as well as uses fewer features to yield better accuracy. These parameters are very important for the NIDS in cloud. Hence our model is better suited for cloud environment.

Table 6.9. Comparison with Earlier Models

Ref. #	Year	Accuracy	True Positive Rate	False Positive Rate	Features
[8]	2018	0.97	0.852	0.210	14
[9]	2019	0.92	0.852	0.214	20
[10]	2020	0.86	0.852	0.214	20
[11]	2020	0.96	0.852	0.211	14
[12]	2020	0.94	0.960	0.360	41
[13]	2021	0.96	0.979	0.076	10
[14]	2021	0.94	0.805	0.214	22
[15]	2022	0.97	0.602	0.214	22
[16]	2023	0.96	0.914	0.214	15
	Proposed Model	**0.98**	**0.987**	**0.021**	7

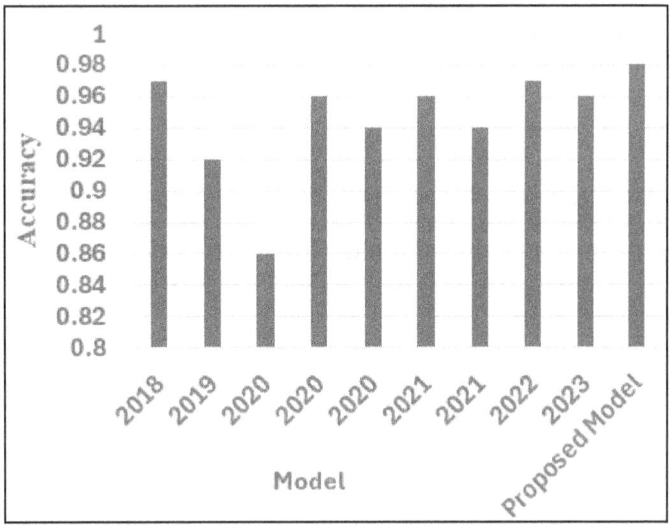

Fig. 6.10. Comparison of Accuracy

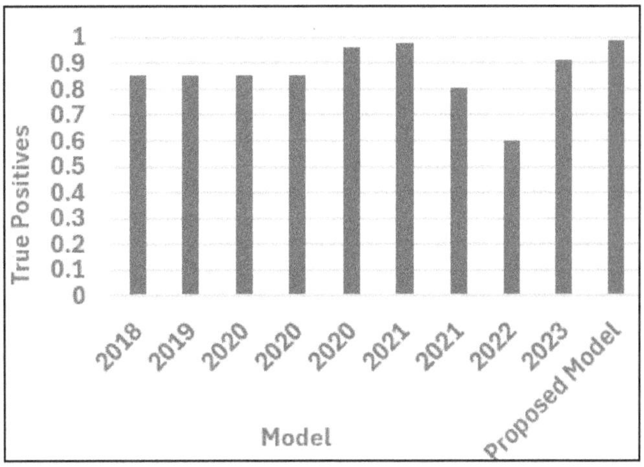

Fig. 6.11. Comparison of True Positive Rate

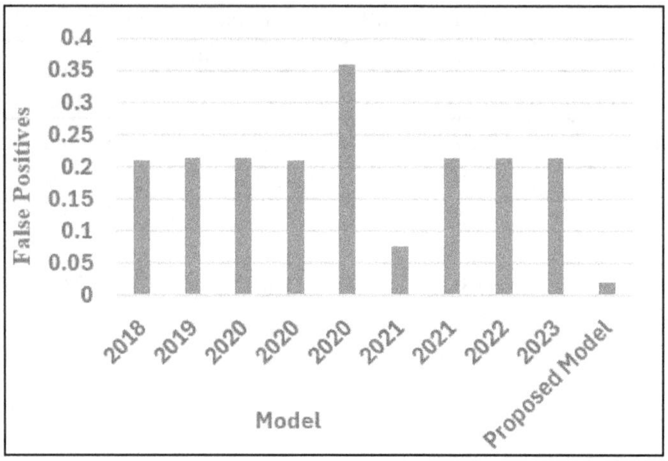

Fig. 6.12. Comparison of False Positive Rate

6.6. Conclusion

With the growing popularity of cloud computing, more and more organizations are migrating their services to the cloud. To ensure the confidentiality, Integrity, and Availability of data, it is important that any intrusion effort is detected in a timely manner to minimize the damage. In this chapter we have proposed an Optimized Collaborative Network Intrusion Detection System in which multiple classifiers at different layers collaborate with each other to increase the accuracy. We have Optimized these classifiers for the improved results. To reduce the dimensionality of data, we have used optimized correlation based feature selection followed by optimized singular vector decomposition. Proposed model achieved a better confusion matrix results as compared to the earlier proposed models.

CHAPTER 7: CONCLUSION AND FUTURE WORK

7.1. Introduction

In this chapter we have provided the concluding remarks about the proposed Collaborative Based Network Intrusion Detection System. We have also discussed about the opportunity for the future work which can optimize the detection rate as well as the number of features of Network Intrusion Detection system for the cloud computing.

7.2. Research Contribution

In this work we proposed Collaborative Based Network Intrusion Detection System using the optimization technique. We used deep neural network and gradient boosted tree as the main classifiers. Correlation based feature selection was optimized using particle swam optimization algorithm. The selected features were further subjected to dimensional reduction using singular vector decomposition method which reduced the number of features to a minimum of seven.

Optimization phase was used to optimize the feature selection, dimensional reduction, and the classifiers. Dimensional reduction was implemented by reducing the number of features using singular vector decomposition and plotting it against the accuracy metric. Similarly, deep neural network and gradient boosted tree were also individually optimized using the training data and evaluating them against accuracy. The optimized configuration of the features and the hyper-parameters of the classifiers were taken as the final configuration. Training was done using 70% of the sample data and testing was done using rest 30%. It was observed that the proposed model performed well on all the performance parameters especially accuracy and the number of features. We were able to reduce the feature by to 7 from the previously improved results pf 15 and at the same time the accuracy of the model was also improved. Keeping in mind the characteristic of cloud computing, the proposed model can be implemented in the cloud environment. The layer-1 and layer-2 model can be used in a distributed computing environment where layer-1 can communicate to layer-2 over the network.

7.3. Future Work

In real life, Intrusion detection datasets are rarely available. Hence, we need to develop network intrusion detection system using normal network traffic only. This is possible using One class classifier. Future work will involve building the similar model using the One class classifier. One class classifier are trained on the instances having normal class only. Such model. when trained can detect the class as "normal" and the instances that do not belong to this class are treated as Intrusions. The main aim should be to eliminate the need of a sample intrusion dataset which is rarely is available. The effort in this case should be to build the model using the sample data which has only "normal' class. Thus, model should be trained on normal data and not the "anomaly" data. The test data should contain both "normal" and "anomaly" classes. This model should be able to identify the maximum instances of "normal" class and then raise alarm for all the remaining instances which will be treated as intrusions. Such a model will be more closely related to the real-world scenarios.

REFERENCES

[1] C. Service and P. Cloud, *Cloud service providers and the cloud ecosystem 2*. 2018. doi: 10.1016/B978-0-12-812810-7.00002-9.

[2] M. A. Khan, “A survey of security issues for cloud computing,” *J. Netw. Comput. Appl.*, vol. 71, pp. 11–29, 2016, doi: 10.1016/j.jnca.2016.05.010.

[3] P. Mell, T. Grance, and T. Grance, “The NIST Definition of Cloud Computing Recommendations of the National Institute of Standards and Technology”.

[4] L. Coppolino, S. D’Antonio, G. Mazzeo, and L. Romano, “Cloud security: Emerging threats and current solutions,” *Comput. Electr. Eng.*, vol. 59, pp. 126–140, 2017, doi: 10.1016/j.compeleceng.2016.03.004.

[5] S. N. Mthunzi, E. Benkhelifa, T. Bosakowski, C. G. Guegan, and M. Barhamgi, “Cloud computing security taxonomy: From an atomistic to a holistic view,” *Futur. Gener. Comput. Syst.*, vol. 107, pp. 620–644, 2020, doi: 10.1016/j.future.2019.11.013.

[6] G. S. Pandi, S. Shah, and K. H. Wandra, “Exploration of Vulnerabilities, Threats and Forensic Issues and its impact on the Distributed Environment of Cloud and its mitigation,” *Procedia Comput. Sci.*, vol. 167, no. 2019, pp. 163–173, 2020, doi: 10.1016/j.procs.2020.03.194.

[7] J. Arshad, P. Townend, and J. Xu, “A novel intrusion severity analysis approach for Clouds,” *Futur. Gener. Comput. Syst.*, vol. 29,

no. 1, pp. 416–428, 2013, doi: 10.1016/j.future.2011.08.009.

[8] A. Carlin, M. Hammoudeh, and O. Aldabbas, "Defence for Distributed Denial of Service Attacks in Cloud Computing," *Procedia Comput. Sci.*, vol. 73, no. Awict, pp. 490–497, 2015, doi: 10.1016/j.procs.2015.12.037.

[9] N. Subramanian and A. Jeyaraj, "Recent security challenges in cloud computing," *Comput. Electr. Eng.*, vol. 71, no. June, pp. 28–42, 2018, doi: 10.1016/j.compeleceng.2018.06.006.

[10] A. Singh and K. Chatterjee, "Cloud security issues and challenges: A survey," *J. Netw. Comput. Appl.*, vol. 79, pp. 88–115, 2017, doi: 10.1016/j.jnca.2016.11.027.

[11] C. Thirumalai, S. Mohan, and G. Srivastava, "An efficient public key secure scheme for cloud and IoT security," *Comput. Commun.*, vol. 150, pp. 634–643, 2020, doi: 10.1016/j.comcom.2019.12.015.

[12] T. Hamed, R. Dara, and S. C. Kremer, *Intrusion Detection in Contemporary Environments*. Elsevier Inc., 2017. doi: 10.1016/b978-0-12-803843-7.00006-5.

[13] A. Tchernykh, U. Schwiegelsohn, E. ghazali Talbi, and M. Babenko, "Towards understanding uncertainty in cloud computing with risks of confidentiality, integrity, and availability," *J. Comput. Sci.*, vol. 36, 2019, doi: 10.1016/j.jocs.2016.11.011.

[14] C. Sauvanaud, M. Kaâniche, K. Kanoun, K. Lazri, and G. Da Silva Silvestre, "Anomaly detection and diagnosis for cloud services: Practical experiments and lessons learned," *J. Syst. Softw.*, vol. 139,

pp. 84–106, 2018, doi: 10.1016/j.jss.2018.01.039.

[15] O. Joldzic, Z. Djuric, and P. Vuletic, "A transparent and scalable anomaly-based DoS detection method," *Comput. Networks*, vol. 104, pp. 27–42, 2016, doi: 10.1016/j.comnet.2016.05.004.

[16] O. Alkadi, N. Moustafa, and B. Turnbull, "A Review of Intrusion Detection and Blockchain Applications in the Cloud: Approaches, Challenges and Solutions," *IEEE Access*, vol. 8, pp. 104893–104917, 2020, doi: 10.1109/ACCESS.2020.2999715.

[17] A. Abid and F. Jemili, "Intrusion Detection based on Graph oriented Big Data Analytics," *Procedia Comput. Sci.*, vol. 176, pp. 572–581, 2020, doi: 10.1016/j.procs.2020.08.059.

[18] S. Iqbal *et al.*, "On cloud security attacks: A taxonomy and intrusion detection and prevention as a service," *J. Netw. Comput. Appl.*, vol. 74, pp. 98–120, 2016, doi: 10.1016/j.jnca.2016.08.016.

[19] C. Modi, D. Patel, B. Borisaniya, H. Patel, A. Patel, and M. Rajarajan, "A survey of intrusion detection techniques in Cloud," *J. Netw. Comput. Appl.*, vol. 36, no. 1, pp. 42–57, 2013, doi: 10.1016/j.jnca.2012.05.003.

[20] S. Rethishkumar and R. Vijayakumar, "State Transition Model (STM): An optimum solution for preventing co-resident DOS attacks in cloud infrastructure," *Mater. Today Proc.*, vol. 27, no. xxxx, pp. 634–639, 2019, doi: 10.1016/j.matpr.2020.01.223.

[21] J. Gu and S. Lu, "An effective intrusion detection approach using SVM with naïve Bayes feature embedding," *Comput. Secur.*, vol.

103, p. 102158, 2021, doi: 10.1016/j.cose.2020.102158.

[22] E. Casey, C. Daywalt, and A. Johnston, *Intrusion Investigation*, 1st ed. Elsevier Inc, 2010. doi: 10.1016/B978-0-12-374267-4.00004-5.

[23] R. Patil, H. Dudeja, and C. Modi, “Designing an efficient security framework for detecting intrusions in virtual network of cloud computing,” *Comput. Secur.*, vol. 85, pp. 402–422, 2019, doi: 10.1016/j.cose.2019.05.016.

[24] S. S. Chakravarthi and R. J. Kannan, “Detection of anomalies in cloud services using network flowdata analysis,” *Int. J. Electr. Eng. Educ.*, 2020, doi: 10.1177/0020720920901436.

[25] H. Jia, J. Liu, M. Zhang, X. He, and W. Sun, “Network intrusion detection based on IE-DBN model,” *Comput. Commun.*, vol. 178, no. July, pp. 131–140, 2021, doi: 10.1016/j.comcom.2021.07.016.

[26] F. J. Abdullayeva, “Distributed denial of service attack detection in E-government cloud via data clustering,” *Array*, vol. 15, no. July, p. 100229, 2022, doi: 10.1016/j.array.2022.100229.

[27] C. Chen *et al.*, “applied sciences FCNN-SE : An Intrusion Detection Model Based on a Fusion CNN and Stacked Ensemble,” 2022.

[28] A. Binbusayyis and T. Vaiyapuri, “Unsupervised deep learning approach for network intrusion detection combining convolutional autoencoder and one-class SVM,” *Appl. Intell.*, vol. 51, no. 10, pp. 7094–7108, 2021, doi: 10.1007/s10489-021-02205-9.

[29] C. F. Tsai and W. C. Lin, “Feature selection and ensemble learning

techniques in one-class classifiers: An empirical study of two-class imbalanced datasets," *IEEE Access*, vol. 9, pp. 13717–13726, 2021, doi: 10.1109/ACCESS.2021.3051969.

[30] E. Mugabo and Q. Zhang, "Intrusion Detection Method Based on Support Vector Machine and Information Gain for Mobile Cloud Computing," vol. 2019, pp. 1–11, 2019.

[31] R. Packer, "Protecting the Network NIDS : the logical first step in intrusion detection deployment," pp. 10–11.

[32] S. Subashini and V. Kavitha, "A survey on security issues in service delivery models of cloud computing," *J. Netw. Comput. Appl.*, vol. 34, no. 1, pp. 1–11, 2011, doi: 10.1016/j.jnca.2010.07.006.

[33] A. Khraisat, I. Gondal, P. Vamplew, and J. Kamruzzaman, "Survey of intrusion detection systems: techniques, datasets and challenges," *Cybersecurity*, vol. 2, no. 1, 2019, doi: 10.1186/s42400-019-0038-7.

[34] M. Wallschläger, A. Gulenko, F. Schmidt, O. Kao, and F. Liu, "Automated Anomaly Detection in Virtualized Services Using Deep Packet Inspection," *Procedia Comput. Sci.*, vol. 110, pp. 510–515, 2017, doi: 10.1016/j.procs.2017.06.137.

[35] P. Mishra, I. Verma, and S. Gupta, "KVMInspector: KVM Based introspection approach to detect malware in cloud environment," *J. Inf. Secur. Appl.*, vol. 51, p. 102460, 2020, doi: 10.1016/j.jisa.2020.102460.

[36] P. Mishra, E. S. Pilli, V. Varadharajan, and U. Tupakula, "Intrusion detection techniques in cloud environment: A survey," *J. Netw.*

Comput. Appl., vol. 77, pp. 18–47, 2017, doi: 10.1016/j.jnca.2016.10.015.

[37] V. Herrera-Semenets, L. Bustio-Martínez, R. Hernández-León, and J. van den Berg, "A multi-measure feature selection algorithm for efficacious intrusion detection," *Knowledge-Based Syst.*, vol. 227, p. 107264, 2021, doi: 10.1016/j.knosys.2021.107264.

[38] W. Fang, X. Tan, and D. Wilbur, "Application of intrusion detection technology in network safety based on machine learning," *Saf. Sci.*, vol. 124, no. December 2019, p. 104604, 2020, doi: 10.1016/j.ssci.2020.104604.

[39] H. Can and Z. Albayrak, "Engineering Science and Technology , an International Journal A hybrid CNN + LSTM-based intrusion detection system for industrial IoT networks," *Eng. Sci. Technol. an Int. J.*, vol. 38, p. 101322, 2023, doi: 10.1016/j.jestch.2022.101322.

[40] T. Saranya, S. Sridevi, C. Deisy, T. D. Chung, and M. K. A. A. Khan, "Performance Analysis of Machine Learning Algorithms in Intrusion Detection System: A Review," *Procedia Comput. Sci.*, vol. 171, no. 2019, pp. 1251–1260, 2020, doi: 10.1016/j.procs.2020.04.133.

[41] I. M. Stephanakis, I. P. Chochliouros, E. Sfakianakis, S. N. Shirazi, and D. Hutchison, "Hybrid self-organizing feature map (SOM) for anomaly detection in cloud infrastructures using granular clustering based upon value-difference metrics," *Inf. Sci. (Ny).*, vol. 494, pp. 247–277, 2019, doi: 10.1016/j.ins.2019.03.069.

[42] J. C. Bezdek, "FCM — the Fuzzy C-Means clustering-algorithm

FCM : THE FUZZY c-MEANS CLUSTERING ALGORITHM," vol. 3004, no. December, 1984, doi: 10.1016/0098-3004(84)90020-7.

[43] A. Ramchandran and A. K. Sangaiah, *Chapter 11 - Unsupervised Anomaly Detection for High Dimensional Data—an Exploratory Analysis*. Elsevier Inc., 2018. doi: 10.1016/B978-0-12-813314-9.00011-6.

[44] N. V Chawla, K. W. Bowyer, L. O. Hall, and W. P. Kegelmeyer, "SMOTE : Synthetic Minority Over-sampling Technique SMOTE : Synthetic Minority Over-sampling Technique," no. June, 2002, doi: 10.1613/jair.953.

[45] N. Bali and A. Singla, "Emerging Trends in Machine Learning to Predict Crop Yield and Study Its Influential Factors: A Survey," *Arch. Comput. Methods Eng.*, vol. 29, no. 1, pp. 95–112, 2022, doi: 10.1007/s11831-021-09569-8.

[46] J. Suto, S. Oniga, and P. P. Sitar, "Comparison of wrapper and filter feature selection algorithms on human activity recognition," *2016 6th Int. Conf. Comput. Commun. Control. ICCCC 2016*, no. Icccc, pp. 124–129, 2016, doi: 10.1109/ICCCC.2016.7496749.

[47] R. V. Deshmukh and K. K. Devadkar, "Understanding DDoS attack & its effect in cloud environment," *Procedia Comput. Sci.*, vol. 49, no. 1, pp. 202–210, 2015, doi: 10.1016/j.procs.2015.04.245.

[48] R. Dechter and J. Pearl, "Generalized Best-First Search Strategies and the Optimality of A," *J. ACM*, vol. 32, no. 3, pp. 505–536, 1985, doi:

10.1145/3828.3830.

[49] T. Fawcett, "An introduction to ROC analysis," *Pattern Recognit. Lett.*, vol. 27, no. 8, pp. 861–874, 2006, doi: 10.1016/j.patrec.2005.10.010.

[50] L. Khatibzadeh, Z. Bornaee, and A. Ghaemi Bafghi, "Applying Catastrophe Theory for Network Anomaly Detection in Cloud Computing Traffic," *Secur. Commun. Networks*, vol. 2019, 2019, doi: 10.1155/2019/5306395.

[51] C. A. de Souza, C. B. Westphall, R. B. Machado, J. B. M. Sobral, and G. dos S. Vieira, "Hybrid approach to intrusion detection in fog-based IoT environments," *Comput. Networks*, vol. 180, 2020, doi: 10.1016/j.comnet.2020.107417.

[52] R. Munir, H. Al-mohannadi, M. Rafiq, W. Campus, W. Cantt, and A. P. Namanya, "Performance Security Trade-off of Network Intrusion Detection and Prevention Systems," no. September, 2016.

[53] C. Security, *Cloud security 11*. 2018. doi: 10.1016/B978-0-12-812810-7.00015-7.

[54] W. Li, W. Meng, and M. H. Au, "Enhancing collaborative intrusion detection via disagreement-based semi-supervised learning in IoT environments," *J. Netw. Comput. Appl.*, vol. 161, no. March, 2020, doi: 10.1016/j.jnca.2020.102631.

[55] Snort Project Team, "SNORT Users Manual 2.9.16," 2020, [Online]. Available: https://www.snort.org/documents

[56] C. N. Modi, D. R. Patel, A. Patel, and M. Rajarajan, "Integrating Signature Apriori based Network Intrusion Detection System (NIDS) in Cloud Computing," *Procedia Technol.*, vol. 6, pp. 905–912, 2012, doi: 10.1016/j.protcy.2012.10.110.

[57] B. Lantz, *Machine Learning with R*, October 20. Birmingham B3 2PB, UK: Packt Publishing Ltd.

[58] N. M. Ibrahim and A. Zainal, "A distributed intrusion detection scheme for cloud computing," *Int. J. Distrib. Syst. Technol.*, vol. 11, no. 1, pp. 68–82, 2020, doi: 10.4018/IJDST.2020010106.

[59] S. Rizvi, G. Labrador, M. Guyan, and J. Savan, "Advocating for Hybrid Intrusion Detection Prevention System and Framework Improvement," *Procedia Comput. Sci.*, vol. 95, pp. 369–374, 2016, doi: 10.1016/j.procs.2016.09.347.

[60] J. Gómez, C. Gil, N. Padilla, R. Baños, and C. Jiménez, "Design of a Snort-Based Hybrid Intrusion," *Proc. 10th Int. Work. Artif. Neural Networks*, pp. 515–522, 2009.

[61] P. K. Rajendran, B. Muthukumar, and G. Nagarajan, "Hybrid intrusion detection system for private cloud: A systematic approach," *Procedia Comput. Sci.*, vol. 48, no. C, pp. 325–329, 2015, doi: 10.1016/j.procs.2015.04.189.

[62] I. Özçelik and R. R. Brooks, "Deceiving entropy based DoS detection," *Comput. Secur.*, vol. 48, pp. 234–245, 2015, doi: 10.1016/j.cose.2014.10.013.

[63] M. A. Aydın, A. H. Zaim, and K. G. Ceylan, "A hybrid intrusion

detection system design for computer network security," *Comput. Electr. Eng.*, vol. 35, no. 3, pp. 517–526, 2009, doi: 10.1016/j.compeleceng.2008.12.005.

[64] M. Uddin, A. A. Rehman, N. Uddin, J. Memon, R. Alsaqour, and S. Kazi, "Signature-based multi-layer distributed intrusion detection system using mobile agents," *Int. J. Netw. Secur.*, vol. 15, no. 2, pp. 97–105, 2013.

[65] G. Kim, S. Lee, and S. Kim, "A novel hybrid intrusion detection method integrating anomaly detection with misuse detection," *Expert Syst. Appl.*, vol. 41, no. 4 PART 2, pp. 1690–1700, 2014, doi: 10.1016/j.eswa.2013.08.066.

[66] Z. Chiba, N. Abghour, K. Moussaid, A. El Omri, and M. Rida, "A Cooperative and Hybrid Network Intrusion Detection Framework in Cloud Computing Based on Snort and Optimized Back Propagation Neural Network," *Procedia Comput. Sci.*, vol. 83, pp. 1200–1206, 2016, doi: 10.1016/j.procs.2016.04.249.

[67] J. Jabez and B. Muthukumar, "Intrusion detection system (ids): Anomaly detection using outlier detection approach," *Procedia Comput. Sci.*, vol. 48, no. C, pp. 338–346, 2015, doi: 10.1016/j.procs.2015.04.191.

[68] S. Teng, N. Wu, H. Zhu, L. Teng, and W. Zhang, "SVM-DT-based adaptive and collaborative intrusion detection," *IEEE/CAA J. Autom. Sin.*, vol. 5, no. 1, pp. 108–118, 2018, doi: 10.1109/JAS.2017.7510730.

[69] A. Javadpour, S. Kazemi Abharian, and G. Wang, “Feature selection and intrusion detection in cloud environment based on machine learning algorithms,” *Proc. - 15th IEEE Int. Symp. Parallel Distrib. Process. with Appl. 16th IEEE Int. Conf. Ubiquitous Comput. Commun. ISPA/IUCC 2017*, pp. 1417–1421, 2018, doi: 10.1109/ISPA/IUCC.2017.00215.

[70] L. Zhang, Z. Y. Bai, Y. L. Lu, Y. X. Zha, and Z. W. Li, “Integrated intrusion detection model based on artificial immune,” *J. China Univ. Posts Telecommun.*, vol. 21, no. 2, pp. 83–90, 2014, doi: 10.1016/S1005-8885(14)60290-9.

[71] N. Afzali Seresht and R. Azmi, “MAIS-IDS: A distributed intrusion detection system using multi-agent AIS approach,” *Eng. Appl. Artif. Intell.*, vol. 35, pp. 286–298, 2014, doi: 10.1016/j.engappai.2014.06.022.

[72] S. Teng, C. Zheng, H. Zhu, D. Liu, and W. Zhang, “A cooperative intrusion detection model for cloud computing networks,” *Int. J. Secur. its Appl.*, vol. 8, no. 3, pp. 107–118, 2014, doi: 10.14257/ijsia.2014.8.3.12.

[73] H. Idrissi, M. Ennahbaoui, E. M. Souidi, and S. El Hajji, “Mobile Agents with Cryptographic Traces for Intrusion Detection in the Cloud Computing,” *Procedia Comput. Sci.*, vol. 73, no. Awict, pp. 179–186, 2015, doi: 10.1016/j.procs.2015.12.010.

[74] T. Daniya, K. Suresh Kumar, B. Santhosh Kumar, and C. Sekhar Kolli, “A survey on anomaly based intrusion detection system,” *Mater. Today Proc.*, no. xxxx, 2021, doi:

10.1016/j.matpr.2021.03.353.

[75] H. Toumi, F. Z. Fagroud, A. Zakouni, and M. Talea, “Implementing hy-ids, mobiles agents and virtual firewall to enhance the security in iaas cloud,” *Procedia Comput. Sci.*, vol. 160, pp. 819–824, 2019, doi: 10.1016/j.procs.2019.11.005.

[76] G. Folino and P. Sabatino, “Ensemble based collaborative and distributed intrusion detection systems: A survey,” *J. Netw. Comput. Appl.*, vol. 66, pp. 1–16, 2016, doi: 10.1016/j.jnca.2016.03.011.

[77] O. Osanaiye, K. K. R. Choo, and M. Dlodlo, “Distributed denial of service (DDoS) resilience in cloud: Review and conceptual cloud DDoS mitigation framework,” *J. Netw. Comput. Appl.*, vol. 67, pp. 147–165, 2016, doi: 10.1016/j.jnca.2016.01.001.

[78] Á. MacDermott, Q. Shi, and K. Kifayat, “Collaborative Intrusion Detection in Federated Cloud Environments,” *J. Comput. Sci. Appl. Big Data Anal. Intell. Syst.*, vol. 3, no. 3, pp. 10–20, 2015, doi: 10.12691/jcsa-3-3A-2.

[79] H. Gupta and V. K. Sharma, “Role of Multiple Encryption in Secure Electronic Transaction,” *Int. J. Netw. Secur. Its Appl.*, vol. 3, no. 6, pp. 89–96, 2011, doi: 10.5121/ijnsa.2011.3606.

[80] N. D. M. and E.-N. Huh, “A Collaborative Intrusion Detection System Framework for Cloud ComputingNo Title,” in *Proceedings of the International Conference on IT Convergence and Security 2011*, 2012, p. 19.

[81] C. J. Fung and Q. Zhu, “FACID: A trust-based collaborative decision

framework for intrusion detection networks," *Ad Hoc Networks*, vol. 53, pp. 17–31, 2016, doi: 10.1016/j.adhoc.2016.08.014.

[82] M. Ficco, L. Tasquier, and R. Aversa, "Intrusion detection in federated clouds," *Int. J. Comput. Sci. Eng.*, vol. 13, no. 3, pp. 219–232, 2016, doi: 10.1504/IJCSE.2016.078929.

[83] M. Idhammad, K. Afdel, and M. Belouch, "Distributed intrusion detection system for cloud environments based on data mining techniques," *Procedia Comput. Sci.*, vol. 127, pp. 35–41, 2018, doi: 10.1016/j.procs.2018.01.095.

[84] A. Abusitta, M. Bellaiche, M. Dagenais, and T. Halabi, "A deep learning approach for proactive multi-cloud cooperative intrusion detection system," *Futur. Gener. Comput. Syst.*, vol. 98, pp. 308–318, 2019, doi: 10.1016/j.future.2019.03.043.

[85] Z. Zhang, J. Wen, J. Zhang, X. Cai, and L. Xie, "A Many Objective-Based Feature Selection Model for Anomaly Detection in Cloud Environment," *IEEE Access*, vol. 8, pp. 60218–60231, 2020, doi: 10.1109/ACCESS.2020.2981373.

[86] J. K. Samriya and N. Kumar, "A novel intrusion detection system using hybrid clustering-optimization approach in cloud computing," *Mater. Today Proc.*, no. xxxx, 2020, doi: 10.1016/j.matpr.2020.09.614.

[87] M. Rabbani, Y. L. Wang, R. Khoshkangini, H. Jelodar, R. Zhao, and P. Hu, "A hybrid machine learning approach for malicious behaviour detection and recognition in cloud computing," *J. Netw. Comput.*

Appl., vol. 151, no. May 2019, p. 102507, 2020, doi: 10.1016/j.jnca.2019.102507.

[88] S. Krishnaveni, S. Sivamohan, S. S. Sridhar, and S. Prabakaran, "Efficient feature selection and classification through ensemble method for network intrusion detection on cloud computing," *Cluster Comput.*, vol. 24, no. 3, pp. 1761–1779, 2021, doi: 10.1007/s10586-020-03222-y.

[89] T. Thilagam and R. Aruna, "Intrusion detection for network based cloud computing by custom RC-NN and optimization," *ICT Express*, vol. 7, no. 4, pp. 512–520, 2021, doi: 10.1016/j.icte.2021.04.006.

[90] M. Mayuranathan, S. K. Saravanan, B. Muthusenthil, and A. Samydurai, "Advances in Engineering Software An efficient optimal security system for intrusion detection in cloud computing environment using hybrid deep learning technique," *Adv. Eng. Softw.*, vol. 173, no. July, p. 103236, 2022, doi: 10.1016/j.advengsoft.2022.103236.

[91] B. Kaushik, R. Sharma, K. Dhama, and A. Chadha, "Performance evaluation of learning models for intrusion detection system using feature selection Performance evaluation of learning models for intrusion detection," *J. Comput. Virol. Hacking Tech.*, no. May, 2023, doi: 10.1007/s11416-022-00460-z.

[92] Z. Mahmood and C. Agrawal, "Intrusion Detection in Cloud Computing environment using Neural Network," *Int. J. ...*, no. September, pp. 1–4, 2012, [Online]. Available: http://ijrcee.org/index.php/ijrcee/article/view/21

[93] M. Ibrahim and R. Elhafiz, "Modeling an intrusion detection using recurrent neural networks," *J. Eng. Res.*, vol. 11, no. 1, p. 100013, 2023, doi: 10.1016/j.jer.2023.100013.

[94] L. Bao, Q. Li, P. Lu, J. Lu, T. Ruan, and K. Zhang, "Execution anomaly detection in large-scale systems through console log analysis," *J. Syst. Softw.*, vol. 143, pp. 172–186, 2018, doi: 10.1016/j.jss.2018.05.016.

[95] F. Nargesian, H. Samulowitz, U. Khurana, E. B. Khalil, and D. Turaga, "Learning Feature Engineering for Classification," no. November, 2017, doi: 10.24963/ijcai.2017/352.

[96] H. Ren, Z. Ye, and Z. Li, "Anomaly detection based on a dynamic Markov model," *Inf. Sci. (Ny).*, vol. 411, pp. 52–65, 2017, doi: 10.1016/j.ins.2017.05.021.

[97] M. Ahmed, A. Naser Mahmood, and J. Hu, "A survey of network anomaly detection techniques," *J. Netw. Comput. Appl.*, vol. 60, pp. 19–31, 2016, doi: 10.1016/j.jnca.2015.11.016.

[98] M. Zhou, L. Han, H. Lu, and C. Fu, "Distributed collaborative intrusion detection system for vehicular Ad Hoc networks based on invariant," *Comput. Networks*, vol. 172, no. January, p. 107174, 2020, doi: 10.1016/j.comnet.2020.107174.

[99] A. Meryem and B. EL Ouahidi, "Hybrid intrusion detection system using machine learning," *Netw. Secur.*, vol. 2020, no. 5, pp. 8–19, 2020, doi: 10.1016/S1353-4858(20)30056-8.

[100] R. Poli, J. Kennedy, and T. Blackwell, "Quantification & Assessment

of the chemical form of residual gadolinium in the brain.pdf," *Swarm Intell*, vol. 1, pp. 33–57, 2007, doi: 10.1007/s11721-007-0002-0.

[101] M. Ihsan Jambak and A. Ikrom Izzuddin Jambak, "Comparison of dimensional reduction using the Singular Value Decomposition Algorithm and the Self Organizing Map Algorithm in clustering result of text documents," *IOP Conf. Ser. Mater. Sci. Eng.*, vol. 551, no. 1, pp. 6–12, 2019, doi: 10.1088/1757-899X/551/1/012046.

[102] J. O. Mebawondu, O. D. Alowolodu, J. O. Mebawondu, and A. O. Adetunmbi, "Network intrusion detection system using supervised learning paradigm," *Sci. African*, vol. 9, 2020, doi: 10.1016/j.sciaf.2020.e00497.

[103] H. M. E. L. Masry, A. E. Khedr, and H. M. Abdul-Kader, "Challenges and opportunities for intrusion detection system in cloud computing environment," *J. Theor. Appl. Inf. Technol.*, vol. 98, no. 20, pp. 3112–3129, 2020.

[104] D. Hazer-rau, R. Arends, and L. Zhang, "Feature Selection Based on Evolutionary Algorithms for Affective Computing and Stress Recognition †," pp. 1–6, 2021.

[105] Bret Lantz, *Machine Learning with R.* Birmingham B3 2PB, UK: Packt Publishing Ltd, 2013. doi: 10.1007/978-981-10-6808-9.

[106] M. K. Dharani, R. Thamilselvan, P. Natesan, P. C. D. Kalaivaani, and S. Santhoshkumar, "Review on Crop Prediction Using Deep Learning Techniques," *J. Phys. Conf. Ser.*, vol. 1767, no. 1, 2021, doi: 10.1088/1742-6596/1767/1/012026.

[107] T. Huang, Y. Zhu, Y. Wu, S. Bressan, and G. Dobbie, "Anomaly detection and identification scheme for VM live migration in cloud infrastructure," *Futur. Gener. Comput. Syst.*, vol. 56, pp. 736–745, 2016, doi: 10.1016/j.future.2015.06.005.

[108] B. Hajimirzaei and N. J. Navimipour, "Intrusion detection for cloud computing using neural networks and artificial bee colony optimization algorithm," *ICT Express*, vol. 5, no. 1, pp. 56–59, 2019, doi: 10.1016/j.icte.2018.01.014.

[109] A. N. Jaber and S. U. Rehman, "FCM–SVM based intrusion detection system for cloud computing environment," *Cluster Comput.*, vol. 23, no. 4, pp. 3221–3231, 2020, doi: 10.1007/s10586-020-03082-6.

[110] W. Xiong *et al.*, "Anomaly secure detection methods by analyzing dynamic characteristics of the network traffic in cloud communications," *Inf. Sci. (Ny).*, vol. 258, no. April, pp. 403–415, 2014, doi: 10.1016/j.ins.2013.04.009.

[111] V. Hnamte and J. Hussain, "Telematics and Informatics Reports DCNNBiLSTM : An Efficient Hybrid Deep Learning-Based Intrusion Detection System," *Telemat. Informatics Reports*, vol. 10, no. November 2022, p. 100053, 2023, doi: 10.1016/j.teler.2023.100053.

[112] K. Gavahi, P. Abbaszadeh, and H. Moradkhani, "DeepYield: A combined convolutional neural network with long short-term memory for crop yield forecasting," *Expert Syst. Appl.*, vol. 184, no. May, 2021, doi: 10.1016/j.eswa.2021.115511.

[113] H. Sharma, G. Hazrati, and J. C. Bansal, "Spider Monkey Optimization Algorithm Spider Monkey Optimization Algorithm," no. January, 2019, doi: 10.1007/978-3-319-91341-4.

[114] W. Song, W. Dong, and L. Kang, "Group anomaly detection based on Bayesian framework with genetic algorithm," *Inf. Sci. (Ny).*, vol. 533, pp. 138–149, 2020, doi: 10.1016/j.ins.2020.03.110.

[115] T. Edition, *Data Mining Practical Machine Learning Tools and Techniques*. AMSTERDAM.

[116] J. C. Bansal, H. Sharma, S. S. Jadon, and M. Gwalior, "Artificial bee colony algorithm : A survey Artificial bee colony algorithm : a survey Jagdish Chand Bansal Harish Sharma * and Shimpi Singh Jadon," no. January, 2013, doi: 10.1504/IJAIP.2013.054681.

www.ingramcontent.com/pod-product-compliance
Lightning Source LLC
LaVergne TN
LVHW082349200926
848481LV00012B/3503